WATER SURVIVAL BIBLE FOR PREPPER'S

How to Locate, Clean, Filtered, and Secured Storage procedures to reduced costs and Ensure Self-reliance

ARIA .S. EDWARD

Copyright © 2024 by [ARIA .S. EDWARD]

Reservation of rights. Except for brief quotations included in critical reviews and certain other non-commercial uses allowed by copyright law, no part of this publication may be duplicated, distributed, or transmitted in any way, including by photocopying, recording, or other electronic or mechanical methods, without the prior written consent of the copyright holder.

TABLE OF CONTENT

TABLE OF CONTENT	3
INTRODUCTION	7
Understanding the benefits of Water in Survival	10
The Role of Preparedness in Water Security	13
CHAPTER ONE	17
Assessing Water Needs	17
Calculating Daily Water Requirements	17
Factors Affecting Water Consumption	20
Planning for Long-Term Water Needs	23
CHAPTER TWO	29
Water Sources	29
Identifying Safe Water Sources in Various Environments	29
Assessing Natural Water Sources (Rivers, Lakes, Streams)	33
Rainwater Harvesting Techniques	37
Well water and groundwater consideration	42
Urban Water Sources and Risks	46
CHAPTER THREE	51
Water Purification Methods	51
Understanding Water Contaminants and Risks	51
Filtration Techniques and Equipment	55

Chemical Treatment Methods (Chlorination, Iodine)	58
UV Sterilization and Solar Disinfection	61
Boiling Water Safely	64

CHAPTER FOUR — 69

Water Storage Strategies	69
Selecting and Preparing Water Containers	69
Long-Term Water Storage Solutions	72
Rotating Water Supplies	76
Creative Storage Options for Limited Spaces	79

CHAPTER FIVE — 83

Water Conservation	83
Importance of Water Conservation in Survival Scenarios	83
Strategies for Reducing Water Usage	86
Repurposing Water for Multiple Uses	89
Hygiene and Personal Care Tips	92

CHAPTER SIX — 95

Emergency Water Procurement	95
Short-Term Solutions for Water Acquisition	95
Improvising Water Collection and Filtration Devices	96
Wilderness Survival Techniques for Finding Water	101

CHAPTER SEVEN — 105

Maintaining Water Quality	105
Regular Inspection and Maintenance of Water Storage	105
Testing Water Quality and Purity	109

Treating Waterborne Illnesses	113
CHAPTER EIGHT	**117**
Community Water Resources	117
Establishing Community Water Plans	117
Cooperation and Collaboration in Water Security	121
Managing Shared Water Resources in Group Settings	125
CHAPTER NINE	**131**
Advanced Water Skills	131
Water Purification from Non-Traditional Sources (Sea Water, Urine)	133
Water Recycling Advanced Filtration and Purification Techniques	137
Water recycling and Reclamation Methods	139
CHAPTER TEN	**145**
Planning for Long-Term Water Security	**145**
Developing Sustainable Water Solutions	145
Permaculture Techniques for Water Management	148
Integration of Water Security into Overall Preparedness Plans	152
CONCLUSION	**156**
Final Thoughts on Water Survival Preparedness	156
Encouragement for Ongoing Learning and adaptation	158

INTRODUCTION

In a world of uncertainty, where natural disasters, emergencies, and unforeseen circumstances can disrupt the availability of essential resources, preparedness becomes paramount. Among the most critical resources for human survival, water reigns supreme. The Prepper's Water Survival Guide serves as a comprehensive manual for individuals and families seeking to safeguard their hydration needs in any situation.

Understanding the Imperative of Water Preparedness

Water, the elixir of life, is not merely a beverage; it is the essence of survival. In emergencies, the availability of clean, potable water can mean the difference between life and death. Yet, the modern world often takes water for granted, assuming an uninterrupted flow from taps and faucets. The truth, however, is far more sobering.

Natural disasters such as earthquakes, hurricanes, and floods can disrupt water infrastructure, leaving entire communities without access to safe drinking water. Even in less catastrophic scenarios, disruptions in municipal water supplies can occur, whether due to contamination, infrastructure failures, or other unforeseen events. For those venturing into the wilderness, remote locations offer no guarantees of freshwater sources, making water scarcity a constant concern.

The Purpose of the Guide

The Prepper's Water Survival Guide is not merely a compendium of facts and figures but a lifeline in times of crisis. It is a roadmap to resilience, offering practical strategies, expert advice, and step-by-step instructions for ensuring water security in any situation. Whether preparing for short-term disruptions or long-term survival scenarios, this guide equips readers with the knowledge, skills, and resources needed to thrive in challenging environments.

Key Themes and Topics

Throughout the guide, several key themes emerge, each addressing critical aspects of water survival preparedness:

Assessing Water Needs: Understanding the daily requirements for hydration and hygiene and planning accordingly.

Identifying Water Sources: Recognizing potential water sources in both urban and wilderness settings and assessing their safety.

Purification Techniques: Learning effective methods for removing contaminants and pathogens from water to ensure its safety for consumption.

Storage Strategies: Implementing proper storage solutions to preserve water supplies and minimize wastage.

Conservation Practices: Adopting habits and techniques to conserve water resources and maximize their utility.

Emergency Procurement: Knowing how to locate, collect, and treat water in emergency situations when standard sources are unavailable.

Community Collaboration: Building networks and partnerships to share resources and support collective water security efforts.

Empowering Individuals for Resilience

Ultimately, the Prepper's Water Survival Guide is about empowerment. It empowers individuals and families to take control of their water destiny, to transform uncertainty into confidence, and to face the future with resilience and resolve. By arming readers with knowledge, skills, and a proactive mindset, this guide serves as a beacon of hope in times of adversity.

As you embark on your journey through these pages, remember that water survival is not just a skill; it is a mindset—a commitment to preparedness, adaptability, and self-reliance. May this guide be your trusted companion on the path to water security, guiding you through the challenges and uncertainties that lie ahead.

Understanding the benefits of Water in Survival

Understanding the importance of water in survival is foundational knowledge for anyone preparing for emergency situations or wilderness expeditions. Water is a fundamental necessity for human life, with the human body being composed of approximately 60% water. In survival scenarios, maintaining proper hydration is crucial for sustaining bodily functions and ensuring overall health. Here's an extensive exploration of the significance of water in survival:

1. Essential Functions of Water in the Body:

Hydration: Water is vital for maintaining the balance of bodily fluids, which play a key role in various physiological processes such as digestion, circulation, and temperature regulation.

Nutrient Transport: Water serves as a medium for transporting essential nutrients, minerals, and oxygen to cells throughout the body.

Temperature Regulation: Sweating, a mechanism for cooling the body, relies on water. Heat-related illnesses such as heat stroke is prevented by adequate hydration which aids normalize body temperature

Waste Removal: Water is necessary for the proper function of the kidneys, which filter waste products from the blood and excrete them through urine. Adequate hydration facilitates the removal of toxins and metabolic waste from the body.

2. **Impact of Dehydration on Health**:
   Physical Performance: Even mild dehydration can impair physical performance, leading to fatigue, weakness, and decreased endurance. In survival situations, maintaining hydration is crucial for sustaining energy levels and optimizing physical capabilities.
   Cognitive Function: Dehydration can negatively impact cognitive function, causing difficulties in concentration, memory, and decision-making. In survival scenarios, clear thinking and mental alertness are essential for problem-solving and decision-making.
   Mood and Emotions: Dehydration can affect mood and emotions, leading to irritability, anxiety, and decreased resilience to stress. In survival situations, maintaining emotional stability is essential for coping with adversity and making sound decisions.

3. **Water Loss and Replacement:**
   Daily Water Requirements: The human body loses water through various processes such as breathing, sweating, urination, and bowel movements. It's essential to replenish lost fluids by consuming an adequate amount of water each day.
   Factors Affecting Water Loss: Environmental conditions, physical activity levels, and individual factors such as age, gender, and health status can influence the rate of water loss. In survival situations, these factors must be taken into account when determining water needs.

4. Challenges of Water Scarcity in Survival Scenarios:

Limited Access to Safe Water Sources: In emergency situations or wilderness environments, access to safe drinking water may be limited or unavailable. Contaminated water sources pose risks of waterborne illnesses such as diarrhea, cholera, and dysentery.

Importance of Water Prioritization: In survival scenarios, water must be prioritized for drinking and essential hygiene needs. Conservation measures may be necessary to stretch limited water supplies until additional sources can be secured.

5. Proactive Water Preparedness:

Planning and Preparedness: Understanding the importance of water in survival underscores the need for proactive planning and preparedness. Preppers and outdoor enthusiasts should have strategies in place for sourcing, purifying, storing, and conserving water in various scenarios.

Skills and Knowledge: Acquiring knowledge and skills related to water purification, navigation, and wilderness survival enhances one's ability to cope with water-related challenges in survival situations.

Equipment and Resources: Having appropriate equipment such as water filters, purification tablets, and portable water containers can significantly improve one's ability to procure and manage water resources in emergencies.

In summary, water is essential for survival, with profound implications for physical health, cognitive function, and emotional well-being. Understanding the importance of water underscores the need for proactive preparedness measures to ensure access to safe drinking water in various survival scenarios. By prioritizing hydration, acquiring relevant skills and knowledge, and maintaining essential equipment, individuals can enhance their resilience and survival capabilities in challenging environments.

The Role of Preparedness in Water Security

In the realm of survivalism, preparedness is not merely a buzzword; it is a way of life—an ethos that guides individuals and communities toward self-reliance and resilience. When it comes to water security, preparedness plays a pivotal role in ensuring survival, especially in the face of uncertainty and adversity. Here's an in-depth exploration of how preparedness enhances water security for preppers:

1. Anticipating and Mitigating Risks:
   Preparedness involves foresight—anticipating potential risks and taking proactive measures to mitigate them. Preppers assess various scenarios, from natural disasters to infrastructure failures, and

identify vulnerabilities in water supply systems. By understanding potential threats, preppers can implement strategies to safeguard water sources, such as installing backup purification systems or reinforcing water storage facilities.

2. Building Redundancy into Water Systems:

Redundancy is a key principle of preparedness, ensuring that multiple layers of protection are in place to maintain water security. Preppers don't rely solely on one water source or purification method; instead, they build redundancy into their systems. This may involve diversifying water sources, investing in multiple filtration technologies, or maintaining backup supplies of potable water.

3. Developing Skills and Knowledge:

Preparedness goes beyond stockpiling supplies; it encompasses the acquisition of skills and knowledge essential for survival. Preppers educate themselves on water purification techniques, navigation strategies, and wilderness survival skills. By developing these competencies, preppers can effectively procure, treat, and conserve water in diverse environments, from urban landscapes to remote wilderness areas.

4. Establishing Contingency Plans:

Preparedness involves contingency planning—developing actionable strategies for responding to emergencies. Preppers create contingency plans specific to water security, outlining steps to take in the event of water shortages, contamination incidents, or supply disruptions. These plans may include alternative water sources, evacuation routes, and communication protocols to coordinate with fellow preppers or emergency responders.

5. Investing in Infrastructure and Equipment:

Preparedness requires investment in infrastructure and equipment designed to enhance water security. Preppers acquire water storage containers, filtration systems, purification tablets, and other tools essential for treating and storing water. They also invest in renewable energy sources, such as solar-powered pumps or portable water distillation units, to ensure continuity of water supply in off-grid scenarios.

6. Promoting Community Resilience:

Preparedness extends beyond individual efforts to encompass community resilience. Preppers recognize the value of collaboration and mutual support in enhancing water security for all members of their community. They engage in community preparedness initiatives, such as neighborhood water-sharing agreements, collaborative water storage projects, or training sessions on water treatment and conservation.

7. Continual Learning and Adaptation:

Preparedness is an ongoing journey, requiring continual learning and adaptation to evolving threats and challenges. Preppers stay informed about emerging risks, technological advancements, and best practices in water security. They remain flexible and adaptable, adjusting their preparedness plans and strategies in response to new information or changing circumstances.

In summary, preparedness is the cornerstone of water security for preppers survival. By anticipating risks, building redundancy, developing skills, establishing contingency plans, investing in infrastructure, promoting community resilience, and embracing a mindset of continual learning and adaptation, preppers enhance their ability to secure and maintain access to clean, potable water in any situation. Preparedness empowers preppers to weather the storms of uncertainty with confidence, resilience, and self-reliance.

CHAPTER ONE

Assessing Water Needs

Calculating Daily Water Requirements

Calculating daily water requirements is essential for maintaining hydration and supporting optimal bodily functions. Here's an extensive guide focusing solely on the calculation process:

1. Baseline Water Needs:
   Start by determining the baseline water needs, which represent the minimum amount of water required to sustain basic bodily functions. This includes water lost through processes such as breathing, sweating, urination, and bowel movements.

2. Consider Individual Factors:
   Take into account individual factors that influence water requirements, such as age, sex, body weight, physical activity level, climate, and health status. Each of these factors can affect how much water a person needs to consume daily.

3. Use General Guidelines:
   Refer to general guidelines provided by health authorities to estimate daily water intake requirements. For instance, the National

Academies of Sciences and Engineering, as well as Medicine recommended:

Approximately 3.7 liters (125 ounces) of total water per day for adult men.

Approximately 2.7 liters (91 ounces) of total water per day for adult women.

4. Adjust for Activity Level:
Adjust water intake based on physical activity level. Individuals engaged in moderate to vigorous activities or those working in hot, humid environments will need to consume more water to replace fluids lost through sweating.

5. Adjust for Climate and Environment:
Take into account the climate and environmental conditions. In hot and dry climates, or at high altitudes, individuals may lose more fluids through evaporation and perspiration, requiring higher water intake to prevent dehydration.

6. Calculate Water Needs:
Use a simple formula to calculate water needs based on body weight. A commonly used formula is to consume 30-35 milliliters (or about 1-1.2 ounces) of water per kilogram of body weight. For example:

A person weighing 70 kilograms (154 pounds) would need approximately 2,100 to 2,450 milliliters (or 70 to 83 ounces) of water per day.

7. Monitor Hydration Status:
Pay attention to signs of dehydration, such as thirst, dark urine, dry mouth, fatigue, and dizziness.

Adjust water intake accordingly to maintain hydration status.

8. Consider Special Circumstances:

Take special circumstances into account when calculating water needs. Pregnant or breastfeeding women, athletes, individuals with specific medical conditions, and those living in extreme environments may have higher water requirements and should adjust accordingly.

9. Adjust for Fluids from Food:

Remember to consider fluids obtained from food sources, as they contribute to overall water intake. Fruits, vegetables, soups, and other foods contain varying amounts of water and can contribute significantly to daily hydration.

10. Use Tools and Resources:

Utilize online calculators, hydration apps, or consult with healthcare professionals to determine individualized water requirements based on specific factors and circumstances.

By following these steps and considering individual factors, it's possible to calculate daily water requirements accurately. Maintaining proper hydration is essential for overall health and well-being, making it crucial to ensure adequate water intake each day.

Factors Affecting Water Consumption

Factors affecting water consumption vary widely and play a significant role in determining how much water individuals need to consume daily, especially for preppers preparing for survival scenarios. Understanding these factors is essential for accurately assessing water needs and planning accordingly. Here's a detailed exploration of the various factors affecting water consumption for preppers:

1. Age:

Age influences water consumption, with specific age groups requiring different amounts of water. Infants, children, and the elderly typically have higher water needs relative to body weight compared to adults.

2. Body Weight:

Body weight directly correlates with water requirements. Generally, larger individuals have higher metabolic rates and greater body mass, necessitating more water to maintain hydration.

3. Sex:

Sex can impact water consumption, with men typically requiring more water than women due to differences in body composition, metabolic rate, and sweat rate.

4. Physical Activity Level:

Physical activity level significantly affects water consumption. Individuals engaged in moderate to vigorous activities or strenuous exercise lose more fluids through sweating and respiration, necessitating increased water intake to replace lost fluids.

5. Environmental Conditions:

Climate, temperature, humidity, and altitude influence water consumption. Hot and humid environments, as well as high altitudes, increase fluid loss through perspiration and evaporation, requiring higher water intake to prevent dehydration.

6. Sweat Rate:

Individuals with higher sweat rates, such as athletes or those working in hot environments, require more water to replace fluids lost through sweating and maintain hydration.

7. Humidity Levels:

High humidity levels can impair the body's ability to cool itself through sweat evaporation, leading to increased fluid loss and higher water requirements.

8. Dietary Factors:

Diet can impact water consumption, with foods containing varying amounts of water contributing to overall hydration. Diets high in fruits, vegetables, and soups provide additional fluids, reducing the need for additional water intake.

9. Caffeine and Alcohol Consumption:

Caffeine and alcohol have diuretic effects, increasing urine output and potentially leading to dehydration. Individuals consuming caffeinated or alcoholic beverages may need to increase their water intake to compensate for fluid losses.

10. Health Status:

Certain medical conditions, such as fever, diarrhea, vomiting, and kidney disease, can increase fluid losses and necessitate higher water intake to maintain hydration. Individuals with medical conditions should consult healthcare professionals to determine appropriate water intake levels.

11. Pregnancy and Breastfeeding:

Pregnant and breastfeeding women have higher water requirements to support fetal development, milk production, and hydration. They may need to increase their water intake to meet these additional needs.

12. Medications:

Certain medications, such as diuretics and antihistamines, can increase urine output and affect fluid balance, necessitating adjustments in water intake to prevent dehydration.

13. Water Quality and Availability:

Access to clean, potable water sources impacts water consumption. In survival scenarios, limited access to safe water may require individuals to ration and conserve water carefully.

Understanding and accounting for these factors is crucial for preppers preparing for survival scenarios. By assessing individual needs and considering environmental conditions, physical activity levels, and other relevant factors, preppers can ensure they have an adequate supply of water to sustain hydration and support overall health and well-being in challenging situations.

Planning for Long-Term Water Needs

In the realm of preparedness, ensuring access to clean water for the long term is paramount. Water is a fundamental resource essential for survival, and preppers must develop comprehensive strategies to meet their water needs over extended periods, especially in scenarios where traditional water sources may become compromised. Here's an extensive guide on planning for long-term water needs for preppers:

1. Assessing Water Requirements:
   Begin by assessing your water requirements for the long term. Consider factors such as the number of individuals in your group, daily water consumption rates, anticipated duration of survival scenarios, and any special considerations such as medical needs or dietary requirements.

2. Establishing Water Storage Systems:

Implement robust water storage systems capable of holding an adequate supply of clean water for an extended period. Choose durable, food-grade containers designed for long-term water storage, such as BPA-free plastic barrels, stainless steel tanks, or collapsible water bladders.

3. Determining Storage Capacity:

Calculate the necessary storage capacity based on your assessed water requirements. Plan for a sufficient supply to cover drinking, cooking, hygiene, and sanitation needs, as well as potential losses due to evaporation, leakage, or contamination.

4. Rotating Water Supplies:

Develop a system for rotating water supplies to ensure freshness and prevent stagnation. Regularly inspect stored water for signs of contamination or degradation, and replace it as needed. Use a first-in, first-out (FIFO) approach to prioritize the consumption of older water supplies.

5. Implementing Water Filtration and Purification Systems:

Invest in reliable water filtration and purification systems capable of treating water from various sources, including surface water, rainwater, and groundwater. Consider portable options such as gravity filters, pump filters, or chemical treatment methods, as well as larger-scale systems for home or retreat use.

6. Exploring Alternative Water Sources:
Identify and explore alternative water sources to supplement stored supplies. These may include rainwater harvesting systems, wells, springs, rivers, or lakes. Evaluate the feasibility and reliability of each source and develop strategies for accessing and treating water as needed.

7. Implementing Rainwater Harvesting Systems:
Install rainwater harvesting systems to capture and store rainwater for long-term use. This may involve collecting rainwater from rooftops using gutters and downspouts, directing it into storage tanks or cisterns, and implementing filtration and treatment systems to ensure water quality.

8. Developing Sustainable Water Solutions:
Consider implementing sustainable water solutions such as permaculture techniques, water-efficient landscaping, and greywater recycling systems. These strategies can reduce reliance on external water sources and promote self-sufficiency over the long term.

9. Integrating Water Security into Overall Preparedness Plans:
Integrate water security into your overall preparedness plans and strategies. Coordinate with other aspects of preparedness, such as food

storage, emergency shelter, and medical supplies, to ensure comprehensive readiness for extended survival scenarios.

10. Community Collaboration and Resource Sharing:

Foster community collaboration and resource sharing to enhance long-term water security. Establish networks with like-minded individuals, neighbors, or community groups to share knowledge, resources, and support in managing water supplies and addressing common challenges.

11. Training and Skill Development:

Invest in training and skill development related to water management, conservation, and treatment. Equip yourself and your group with the knowledge and expertise needed to effectively manage water resources and respond to water-related emergencies in a self-reliant manner.

12. Regular Review and Adaptation:

Continually review and adapt your long-term water planning strategies based on changing circumstances, emerging threats, and lessons learned from real-world experiences. Remain flexible and proactive in adjusting your approach to ensure ongoing water security and resilience.

By diligently planning for long-term water needs, preppers can enhance their ability to withstand and thrive in challenging environments, ensuring access to clean, potable water for themselves and their communities over extended periods of time.

CHAPTER TWO

Water Sources

Identifying Safe Water Sources in Various Environments

Access to safe water is critical for survival in any environment, but identifying reliable sources can be challenging, especially in emergency situations. Preppers must be able to recognize and assess water sources to ensure they are safe for consumption. Here's an extensive guide on identifying safe water sources in various environments for preppers:

1. Rivers and Streams:
   Identification: Rivers and streams are natural watercourses that flow through diverse landscapes, offering potential water sources in wilderness and rural areas.
   Assessment: Choose flowing water over stagnant pools, as it is less likely to be contaminated. Look for clear, clean-looking water with minimal sediment or debris. Avoid water downstream from industrial sites, agricultural areas, or human settlements, as it may be polluted.

2. Lakes and Ponds:
   Identification: Lakes and ponds are large bodies of standing water found in both natural and artificial

environments, providing potential water sources in rural and wilderness settings.

Assessment: Select lakes and ponds with clear, blue-green water indicating minimal contamination. Avoid stagnant, foul-smelling water with visible algae blooms or discoloration, as it may contain harmful pathogens or toxins.

3. Springs and Seeps:

Identification: Springs and seeps are natural sources of groundwater that emerge from the earth's surface, often found in mountainous or forested regions.

Assessment: Look for clear, cold water bubbling up from the ground, indicating a clean and reliable water source. Test the water for purity and ensure there are no signs of contamination from nearby human or animal activity.

4. Rainwater:

Identification: Rainwater is precipitation that falls from the sky and can be collected from rooftops, tarps, or other impervious surfaces.

Assessment: Collect rainwater in clean, food-grade containers to prevent contamination. Use rainwater harvesting techniques to capture and store water for non-potable uses such as irrigation or hygiene. Treat rainwater before drinking to remove impurities and ensure safety.

5. Groundwater Wells:

Identification: Groundwater wells are man-made structures drilled into aquifers beneath the earth's surface, providing access to underground water reserves.

Assessment: Test well water for quality and purity before consumption, as it may contain contaminants such as bacteria, nitrates, or heavy metals. Ensure proper well construction, maintenance, and regular testing to prevent contamination and ensure water safety.

6. Urban Sources:

Identification: Urban environments offer access to municipal water supplies, bottled water, and other commercial sources of potable water.

Assessment: Use municipal water supplies where available, as they are treated and monitored to meet safety standards. Purchase bottled water from reputable brands or retailers to ensure quality and safety. Avoid drinking water from unknown or unverified sources in urban areas.

7. Snow and Ice:

Identification: Snow and ice can be melted to obtain water in cold climates or high-altitude environments.

Assessment: Melt snow or ice in clean containers over a heat source to obtain water. Filter or purify melted snow or ice to remove impurities and ensure safety before drinking.

8. Coastal Sources:

Identification: Coastal environments offer access to seawater, which can be desalinated or filtered to obtain potable water.

Assessment: Use desalination methods such as distillation or reverse osmosis to remove salt and impurities from seawater. Ensure proper equipment and procedures are followed to obtain safe drinking water from coastal sources.

9. Desert Sources:
Identification: Desert environments may contain oases, natural springs, or hidden water sources that can sustain life.

Assessment: Search for signs of vegetation, wildlife, or topographic features indicating the presence of water in desert regions. Use caution when exploring desert sources and ensure water is tested for safety before consumption.

10. Industrial and Agricultural Runoff:
Identification: Industrial and agricultural activities may produce runoff containing pollutants that can contaminate surface water sources.

Assessment: Avoid water sources downstream from industrial facilities, agricultural fields, or livestock operations, as they may be contaminated with chemicals, pesticides, or pathogens. Look for alternative sources of clean water away from areas of potential contamination.

Identifying safe water sources in various environments is essential for preppers preparing for survival scenarios. By understanding the characteristics of different water sources and assessing them for safety, preppers can ensure access to clean, potable water to meet their hydration and survival needs. Incorporate water identification and assessment techniques into your preparedness plans to enhance resilience and self-sufficiency in challenging environments.

Assessing Natural Water Sources (Rivers, Lakes, Streams)

Natural water sources such as rivers, lakes, and streams can be valuable assets for preppers seeking to secure their water supply in survival scenarios. However, it is essential to assess these sources carefully to ensure they are safe for consumption and suitable for survival needs. Here's a detailed guide on assessing natural water sources for preppers:

1. Visual Inspection:

Appearance: Observe the appearance of the water source. Clear, clean-looking water is preferable, while water that appears cloudy, murky, or discolored may indicate contamination.

Debris:Look for signs of debris, sediment, or floating material in the water. Excessive debris may indicate pollution or runoff from nearby areas.

2. Smell and Odor:

Odor: Take note of any unusual or foul odors emanating from the water. Foul-smelling water may indicate the presence of organic matter, algae, or pollutants.

Chemical Smells: Be cautious of chemical odors such as gasoline, sewage, or industrial chemicals, which could indicate contamination.

3. Water Flow:

Flow Rate: Assess the flow rate of rivers, streams, or creeks. Fast-flowing water is less likely to be contaminated than stagnant or slow-moving water.

Stagnation: Avoid stagnant pools or backwaters, as they are more prone to bacterial growth and contamination.

4. Surrounding Environment:

Land Use: Consider the land use surrounding the water source. Avoid areas downstream from industrial facilities, agricultural operations, or urban developments, as they may contribute pollutants to the water.

Vegetation: Look for healthy vegetation along the banks of rivers and streams, as it indicates a natural and less disturbed ecosystem.

5. Wildlife Presence:

Wildlife Activity: Observe wildlife activity around the water source. Healthy populations of fish, birds, and other wildlife can indicate a balanced and thriving ecosystem.

Waterfowl: Be cautious of waterfowl congregating in large numbers, as their presence may contribute to fecal contamination.

6. Water Testing:

Bacterial Testing: Conduct bacterial testing to assess water quality for pathogens such as E. coli, fecal coliforms, and other harmful bacteria. Use water testing kits or portable water quality meters to perform on-site testing.

Chemical Testing: Test water samples for chemical contaminants such as heavy metals, pesticides, and industrial pollutants. Chemical testing may require laboratory analysis for accurate results.

7. Filtration and Treatment:

Filtration: Use portable water filtration devices or methods to remove sediment, debris, and particulate matter from natural water sources. Filtration can improve water clarity and reduce the risk of ingesting contaminants.

Treatment: Consider water treatment methods such as boiling, chemical disinfection, or ultraviolet (UV) sterilization to kill harmful pathogens and make water safe for consumption.

8. Accessibility and Safety:

Accessibility: Evaluate the accessibility of the water source for collection and transportation. Choose locations that are easily accessible and safe to reach, especially in emergency situations.

Safety Precautions: Take necessary safety precautions when accessing natural water sources, such as wearing appropriate footwear, avoiding slippery rocks or banks, and watching out for potential hazards such as swift currents or wildlife.

9. Long-Term Viability:

Sustainability: Consider the long-term viability of the water source for sustained use. Assess factors such as seasonal variations in flow rates, drought conditions, and potential impacts of climate change on water availability.

Backup Sources: Identify alternative water sources in the vicinity as backup options in case the primary water source becomes compromised or unavailable.

10. Documentation and Monitoring:

Record Keeping: Keep detailed records of water assessments, test results, and any observations related to the water source. Documentation can help track changes over time and guide future decision-making.

Regular Monitoring: Conduct regular monitoring and reassessment of natural water sources to ensure ongoing safety and suitability for consumption. Stay vigilant for signs of contamination or environmental changes that may affect water quality.

By following these assessment guidelines, preppers can effectively evaluate natural water sources such as rivers, lakes, and streams to ensure they are safe, reliable, and suitable for meeting their water survival needs in emergency situations. Incorporate these assessment techniques into your preparedness plans to enhance resilience and self-sufficiency in challenging environments.

Rainwater Harvesting Techniques

Rainwater harvesting is a valuable technique for preppers to collect and store water for various uses, especially in situations where access to clean water may be limited. Here's an in-depth exploration of different rainwater harvesting techniques along with explanations for a prepper's water survival guide:

1. Roof-Based Rainwater Harvesting:
   Explanation: This method involves collecting rainwater from the roof of a building, typically using gutters and downspouts to direct water into storage tanks or cisterns.
   Components:
   Gutters: Channels attached to the edges of the roof to collect rainwater and direct it towards downspouts.
   Downspouts: Vertical pipes connected to gutters, carrying rainwater from the roof to storage containers.
   Storage Tanks/Cisterns: Containers used to store collected rainwater for later use.
   Benefits: Roof-based rainwater harvesting is a simple and effective method for capturing rainwater from large surface areas, making it suitable for both urban and rural settings.

2. Ground-Based Rainwater Harvesting:

Explanation: Ground-based rainwater harvesting involves collecting rainwater directly from the ground surface and channeling it into storage reservoirs or infiltration systems.

Components:

Swales: Shallow depressions or ditches designed to capture and direct surface runoff towards storage areas or infiltration basins.

Berms: Raised earth structures constructed to slow down and capture rainwater, allowing it to infiltrate into the soil.

Rain Gardens: Landscaped areas planted with water-absorbing vegetation to capture and retain rainwater.

Benefits: Ground-based rainwater harvesting techniques are suitable for landscapes with permeable soils and can help replenish groundwater reserves while reducing runoff and erosion.

3. Rainwater Collection from Natural Surfaces:

Explanation: This method involves collecting rainwater from natural surfaces such as rock formations, cliffs, or valleys and channeling it into storage containers or impoundments.

Components:

Rock Catchments: Sloping rock surfaces designed to direct rainwater towards collection points or storage areas.

Terraces: Step-like structures built into hillsides or slopes to slow down rainwater runoff and promote infiltration.

Check Dams: Low barriers constructed across streams or drainage channels to capture sediment and increase water retention.

Benefits: Rainwater collection from natural surfaces utilizes existing landforms to capture and store rainwater, making it a sustainable and low-cost option for water harvesting in remote or wilderness areas.

4. Rainwater Harvesting with Rain Barrels:

Explanation: Rain barrels are small-scale storage containers placed directly under downspouts to collect rainwater from rooftops for immediate use.

Components:

Rain Barrels: Large, food-grade barrels or containers equipped with a lid, screen, and overflow valve to collect and store rainwater.

Diverter Kits: Devices installed on downspouts to divert rainwater into rain barrels while preventing debris from entering the storage container.

Overflow Systems: Pipes or hoses attached to rain barrels to divert excess water away from buildings or structures.

Benefits: Rain barrels are affordable, easy to install, and suitable for residential or small-scale water harvesting applications, providing a convenient source of water for gardening, landscaping, or non-potable uses.

5. Portable Rainwater Collection Systems:

Explanation: Portable rainwater collection systems are designed for on-the-go water harvesting, allowing preppers to collect and transport rainwater from various sources to their survival shelters or bug-out locations.

Components:

Collapsible Containers: Lightweight, foldable containers or bags made of durable materials such as PVC or polyethylene, designed for easy transport and storage.

Rainwater Filtration Devices: Portable water filtration systems or purification tablets to treat collected rainwater for drinking or cooking purposes.

Collection Accessories: Tarpaulins, tarps, or ponchos used to collect rainwater in the field and funnel it into portable containers.

Benefits: Portable rainwater collection systems provide preppers with a flexible and mobile water harvesting solution, enabling them to replenish water supplies while on the move or in temporary shelter locations.

6. Rainwater Harvesting from Greenhouses or Polytunnels:

Explanation: Greenhouses or polytunnels can serve as effective rainwater harvesting structures, collecting rainwater from their roofs and directing it into storage tanks or reservoirs for irrigation or other agricultural uses.

Components:
Greenhouse Roofs: Transparent or translucent roofing materials designed to allow sunlight to penetrate while capturing rainwater for collection.
Gutter Systems: Gutters installed along the edges of greenhouse roofs to collect rainwater and channel it into storage containers or irrigation systems.
Irrigation Systems: Drip irrigation or soaker hose systems connected to rainwater storage tanks, delivering water directly to plant roots for efficient water use.
Benefits: Rainwater harvesting from greenhouses or polytunnels provides preppers with a sustainable water source for growing food crops, maintaining greenhouse humidity, and supporting agricultural production in controlled environments.

7. Rainwater Harvesting from Vehicles or Tarps
Explanation: In emergency situations or survival scenarios, preppers can collect rainwater directly from vehicles, tarps, or makeshift shelters using improvised collection methods.
Components:
Tarps or Plastic Sheeting: Waterproof materials spread over surfaces to collect rainwater and funnel it into containers or storage
Vehicle Roofs: Car roofs or truck beds used as rainwater collection surfaces, with gutters or channels directing water into containers or jerry cans.

Containers: Buckets, jugs, or bottles placed under tarps or vehicle roofs to collect rainwater for drinking, cooking, or hygiene purposes.

Benefits: Rainwater harvesting from vehicles or tarps provides preppers with a quick and improvised water collection method in emergency situations, allowing them to replenish water supplies without access to traditional harvesting infrastructure.

By incorporating these rainwater harvesting techniques into their preparedness plans, preppers can ensure access to a reliable and sustainable water source for their survival needs, even in challenging or off-grid environments. Each method offers unique advantages and can be adapted to suit various scenarios, providing preppers with versatile options for securing their water supply in times of need.

Well water and groundwater consideration

Introduction:
Well water and groundwater are valuable resources for preppers seeking to secure their water supply in survival scenarios. Groundwater sources offer reliability and independence from municipal water systems, making them attractive options for self-reliant individuals and communities. Here's an extensive guide on considerations for utilizing well

water and groundwater in preppers' water survival plans:

1. Understanding Well Water:
Well water refers to groundwater accessed through wells drilled into aquifers beneath the earth's surface. Wells draw water from underground reservoirs, providing a natural and often high-quality water source for drinking, cooking, hygiene, and irrigation.

2. Assessing Well Suitability:
Before relying on well water, preppers should assess the suitability of potential well sites. Factors to consider include groundwater availability, depth to the water table, aquifer characteristics, soil composition, geological features, and local regulations governing well construction and usage.

3. Well Construction and Maintenance:
Invest in proper well construction and maintenance to ensure the safety and reliability of well water. Work with licensed professionals to drill wells, install casing, and construct wellheads according to industry standards and regulatory requirements. Regular inspection, testing, and maintenance of wells are essential to prevent contamination and ensure optimal performance.

4. Water Quality Testing:
Conduct comprehensive water quality testing to assess the safety and suitability of well water for

drinking. Test for contaminants such as bacteria, nitrates, heavy metals, pesticides, and volatile organic compounds (VOCs) that may pose health risks. Regular testing is recommended to monitor water quality over time and identify any changes or potential sources of contamination.

5. Water Treatment and Filtration:
Depending on the results of water quality testing, preppers may need to implement water treatment and filtration systems to remove contaminants and ensure the safety of well water. Common treatment methods include chlorination, UV disinfection, reverse osmosis, activated carbon filtration, and ion exchange. Select treatment technologies based on specific water quality issues and individual needs.

6. Storage and Distribution:
Install proper storage and distribution systems to manage well water supply. Use food-grade storage tanks, cisterns, or reservoirs to store collected water, protecting it from contamination and maintaining water quality. Implement distribution networks with pumps, pipes, and valves to deliver water to various points of use throughout the property or survival shelter.

7. Backup Power and Redundancy:
Plan for backup power and redundancy to ensure continuous access to well water, especially during power outages or emergencies. Install backup generators, solar-powered pumps, or

hand-operated pumps to maintain water supply in off-grid scenarios. Consider storing additional water supplies as a backup in case of well failure or equipment malfunction.

8. Sustainable Water Management;
Practice sustainable water management techniques to conserve and protect groundwater resources. Implement water-saving measures such as rainwater harvesting, greywater recycling, and water-efficient landscaping to reduce reliance on well water and minimize environmental impact. Adopt responsible land use practices to prevent groundwater contamination and promote long-term water sustainability.

9. Community Collaboration:
Foster community collaboration and resource sharing to enhance well water resilience and collective water security. Coordinate with neighboring property owners, well users, and local authorities to share knowledge, resources, and best practices for managing well water effectively. Establish mutual aid networks to support each other during water-related emergencies or crises.

10. Emergency Preparedness:
Include well water considerations in overall emergency preparedness planning. Develop contingency plans for addressing potential well failures, water shortages, or contamination events. Stockpile emergency supplies such as water

treatment chemicals, spare parts, and backup equipment to address unforeseen challenges and maintain water supply resilience.

Conclusion:
Well water and groundwater offer preppers a reliable and self-sufficient water source for survival preparedness. By understanding the principles of storage, and sustainable management, preppers can harness the potential of well water to meet their long-term water needs and enhance their resilience in times of uncertainty. Integrating well water considerations into comprehensive water survival plans ensures access to clean, potable water for self-reliant living and emergency preparedness.

Urban Water Sources and Risks

Definition of Urban Water:
Urban water refers to the water resources available within urban areas, including both natural and human-made sources such as municipal water supplies, groundwater, surface water bodies, rainwater, and recycled water. In urban environments, water is essential for drinking, sanitation, hygiene, industrial processes, irrigation, and recreational purposes.

Sources of Urban Water:
1. Municipal Water Supply:
   Municipal water supplies are the primary source of drinking water for urban populations. Water is

sourced from surface water reservoirs (lakes, rivers, and dams) or groundwater aquifers, treated at water treatment plants to meet drinking water standards, and distributed through a network of pipes to homes and businesses.

2. Groundwater:

Groundwater is another vital source of water in urban areas, accessed through wells and boreholes. Groundwater provides a reliable and often high-quality water supply, but overexploitation can lead to depletion of aquifers and saltwater intrusion in coastal areas.

3. Surface Water Bodies:

Surface water bodies such as rivers, lakes, and reservoirs contribute to urban water supplies, serving as sources for drinking water, recreational activities, and wildlife habitat. However, surface water sources are susceptible to pollution from urban runoff, industrial discharges, and sewage effluent.

4. Rainwater:

Rainwater harvesting is increasingly utilized in urban areas to supplement municipal water supplies. Rainwater collected from rooftops or paved surfaces can be stored and treated for non-potable uses such as irrigation, flushing toilets, or washing vehicles.

5. Recycled Water:

Recycled water, also known as reclaimed water or wastewater effluent, is treated sewage water that is purified to meet water quality standards for non-potable uses. Recycled water is used for irrigation, industrial processes, and environmental restoration in urban settings.

Risks Associated with Urban Water Sources:
1. Contamination:

Urban water sources are susceptible to contamination from various sources, including industrial pollutants, agricultural runoff, sewage discharges, and hazardous waste. Contaminated water poses risks of waterborne diseases such as cholera, typhoid, and gastrointestinal illnesses.

2. Infrastructure Failures:

Aging or poorly maintained water infrastructure in urban areas can lead to leaks, breaks, and system failures, resulting in disruptions to water supply and potential contamination of drinking water. Infrastructure failures may occur due to corrosion, pipe bursts, or inadequate maintenance.

3. Chemical Contaminants:

Urban water sources may contain chemical contaminants such as heavy metals, pesticides, pharmaceuticals, and volatile organic compounds (VOCs) from industrial activities, urban runoff, and improper disposal of household chemicals.

Exposure to these contaminants can have adverse health effects over time.

4. Natural Disasters:

Urban areas are vulnerable to natural disasters such as floods, hurricanes, and earthquakes, which can damage water infrastructure, disrupt water supply systems, and contaminate water sources. Preppers must be prepared to secure alternative water sources and treat water for safety in the aftermath of disasters.

5. Water Shortages:

Urban water supplies may face shortages due to increased demand, population growth, climate change, drought, or inadequate water management practices. Water shortages can lead to rationing, restrictions, or conflicts over water allocation, affecting both residents and businesses.

6. Cybersecurity Threats:

Urban water systems are increasingly reliant on digital technologies for monitoring, control, and communication. Cybersecurity threats such as hacking, malware, or ransomware attacks pose risks to water infrastructure, potentially disrupting operations and compromising water quality.

7. Disinfection Byproducts:

Chlorination and other disinfection methods used to treat drinking water can produce disinfection byproducts (DBPs) such as trihalomethanes

(THMs) and haloacetic acids (HAAs), which are known carcinogens. Exposure to DBPs over time may increase the risk of cancer and other health problems.

Understanding the sources and risks associated with urban water is crucial for preppers preparing for survival scenarios. By assessing the vulnerabilities of urban water sources to contamination, infrastructure failures, natural disasters, and other threats, preppers can develop effective strategies to secure alternative water sources, treat water for safety, and ensure access to clean, potable water for their survival needs.

CHAPTER THREE

Water Purification Methods

Understanding Water Contaminants and Risks

Water contamination poses significant risks to preppers relying on natural or stored water sources for survival. Understanding the various contaminants and their associated risks is crucial for ensuring access to safe and potable water in emergency situations. Here's an extensive guide on water contaminants and risks for preppers:

1. Biological Contaminants:
   Bacteria: Common bacterial contaminants include Escherichia coli (E. coli), Salmonella, and Campylobacter. These pathogens can cause gastrointestinal illnesses such as diarrhea, vomiting, and abdominal cramps.
   Viruses: Viral contaminants such as norovirus, rotavirus, and hepatitis A can cause severe gastrointestinal infections and flu-like symptoms.
   Protozoa: Parasitic protozoa such as Giardia and Cryptosporidium can lead to waterborne diseases like giardiasis and cryptosporidiosis, characterized by diarrhea and dehydration.

2. Chemical Contaminants:
   Heavy Metals: Toxic heavy metals such as lead, arsenic, mercury, and cadmium can leach into

water sources from industrial activities, mining operations, and natural deposits. Chronic exposure to heavy metals can cause organ damage, neurological disorders, and developmental issues.

Pesticides and Herbicides: Agricultural chemicals such as pesticides and herbicides may contaminate water sources through runoff or leaching from fields. Exposure to these chemicals can lead to acute poisoning or long-term health effects.

Industrial Pollutants: Industrial pollutants such as polychlorinated biphenyls (PCBs), volatile organic compounds (VOCs), and dioxins can contaminate water sources near manufacturing facilities, waste sites, or industrial zones. These contaminants may cause cancer, reproductive disorders, and other adverse health effects.

Chlorine and Chloramines: Disinfection byproducts such as chlorine and chloramines, formed during water treatment, can pose health risks if present in high concentrations. Long-term exposure to disinfection byproducts may increase the risk of cancer and other health problems.

3. Physical Contaminants:

Sediment and Particulate Matter: Sediment, debris, and particulate matter can enter water sources from erosion, runoff, or human activities. Excessive sedimentation can affect water clarity, clog filtration systems, and harbor pathogens.

Suspended Solids: Suspended solids such as silt, sand, and clay can reduce water quality and interfere with water treatment processes. These particles may carry bacteria, viruses, or chemical contaminants, posing health risks if ingested.

4. Radiological Contaminants:

Radioactive Elements: Radioactive contaminants such as radium, uranium, and radon can enter water sources from natural deposits, mining activities, or nuclear accidents. Prolonged exposure to radioactive elements may increase the risk of cancer and genetic mutations.

5. Emerging Contaminants:

Pharmaceuticals and Personal Care Products: Emerging contaminants such as pharmaceuticals, hormones, and personal care products can enter water sources through sewage effluent or improper disposal. These contaminants may disrupt endocrine function, affect aquatic ecosystems, and pose unknown health risks to humans.

Microplastics: Microplastics, tiny plastic particles less than 5 millimeters in size, have been found in water sources worldwide. Microplastics can absorb and release chemical pollutants, accumulate in the environment, and potentially enter the food chain, posing risks to human health and the environment.

6. Risks Associated with Water Contaminants:

Acute Health Effects: Exposure to water contaminants can cause acute health effects such as gastrointestinal illness, nausea, vomiting, diarrhea, and skin irritation. These effects are particularly severe in vulnerable populations such as children, the elderly, and individuals with weakened immune systems.

Chronic Health Effects: Long-term exposure to certain contaminants may lead to chronic health effects such as cancer, organ damage, neurological disorders, reproductive issues, and developmental abnormalities. Chronic health effects may not manifest immediately but can occur over extended periods of time.

Environmental Impacts: Water contaminants can harm aquatic ecosystems, wildlife, and natural habitats. Pollution of water sources can lead to fish kills, algal blooms, habitat destruction, and ecosystem imbalances, affecting biodiversity and ecosystem services.

Infrastructure Damage: Certain contaminants, such as corrosive chemicals or heavy metals, can damage water infrastructure, pipes, and treatment facilities. Infrastructure damage may lead to water leaks, pipe bursts, and system failures, compromising water quality and availability.

Conclusion:
Understanding water contaminants and associated risks is essential for preppers preparing for survival scenarios. By recognizing the types of contaminants that may be present in water sources and understanding their potential health effects, preppers can take proactive measures to mitigate risks, ensure access to safe drinking water, and safeguard their health and well-being in emergency

situations. Incorporate water testing, filtration, and treatment methods into your preparedness plans to address potential contaminants and ensure the availability of clean, potable water for survival needs.

Filtration Techniques and Equipment

Water filtration is crucial for preppers preparing for survival situations where clean water may be scarce or contaminated. Here's a detailed overview of filtration techniques and equipment commonly used in water purification:

1. Mechanical Filtration:
Description: Mechanical filtration involves physically removing particles from water by passing it through a barrier with small pores.
Equipment:
Filter Straws: Portable and lightweight, filter straws allow users to drink directly from water sources, removing sediment and contaminants.
Pump Filters: Hand-operated pump filters draw water through a filtration element, usually made of ceramic or fiberglass, to remove debris and microbes.
Gravity Filters: Gravity-fed systems consist of two compartments with a filter element in between. Water is poured into the top compartment and gravity pulls it through the filter into the clean water reservoir.

Advantages: Simple to use, effective at removing visible particles and sediment.

Limitations: May not remove microscopic pathogens, requires regular maintenance and replacement of filter elements.

2. Chemical Filtration:

Description: Chemical filtration involves using substances to chemically bind to and neutralize contaminants in water.

Equipment:

Iodine Tablets: Dissolve iodine tablets in water to kill bacteria, viruses, and some parasites.

Chlorine Dioxide Drops: Effective against a wide range of pathogens, chlorine dioxide drops are added to water and left to disinfect for a specified period.

Activated Carbon Filters: Carbon filters absorb organic compounds, chlorine, and other chemicals, improving water taste and odor.

Advantages: Lightweight, compact, effective against a broad range of contaminants.

Limitations: Chemical taste in treated water, not always effective against all types of pathogens, may require long wait times for disinfection.

3. Biological Filtration:

Description: Biological filtration involves using living organisms or biological processes to treat water.

Equipment:
Biofilters: Biofilters contain layers of beneficial bacteria that break down organic matter and remove contaminants from water.
Aquarium Filters: Some aquarium filters, such as sponge filters and canister filters, can be adapted for water purification by promoting the growth of beneficial bacteria.
Advantages: Natural and sustainable method, effective at removing organic pollutants.
Limitations: Requires time for bacteria to establish, may not remove all contaminants, requires regular maintenance.

4. Distillation:
Description: Distillation involves heating water to create steam, which is then cooled and condensed back into liquid form, leaving behind contaminants.
Equipment:
Solar Stills: Solar stills use sunlight to heat water and collect condensed steam as purified water.
Countertop Distillers: Electric countertop distillers heat water, collect steam, and condense it into clean water in a separate container.
Advantages: Removes almost all contaminants, including heavy metals and pathogens.
Limitations: Energy-intensive, slow process, does not remove volatile organic compounds.

5. **Reverse Osmosis**:
Description: Reverse osmosis (RO) involves forcing water through a semipermeable membrane that blocks contaminants, leaving behind clean water.
Equipment:
Reverse Osmosis Systems: These systems typically consist of a pre-filter, a membrane filter, and a post-filter to remove sediment, chlorine, and other contaminants.
Advantages: Removes a wide range of contaminants, including heavy metals, salts, and pathogens.
Limitations: Requires electricity or pressure to operate, produces wastewater, initial investment and maintenance costs can be high.
When preparing for water survival situations, preppers should consider a combination of these filtration techniques and equipment to ensure access to clean and safe drinking water. Regular maintenance, proper storage, and thorough understanding of each filtration method's capabilities are essential for effective water purification.

Chemical Treatment Methods (Chlorination, Iodine)

Chemical treatment methods are crucial for preppers preparing for survival situations where access to clean water may be limited or contaminated. Chlorination and iodine are two commonly used chemical treatment methods for water purification. Here's an extensive overview of each:

1. Chlorination:

Description: Chlorination involves adding chlorine-based compounds to water to disinfect it by killing or inactivating harmful microorganisms such as bacteria, viruses, and protozoa.

Types of Chlorine Compounds:
Chlorine Bleach (Sodium Hypochlorite)**: Household bleach containing sodium hypochlorite can be used to disinfect water. It typically contains 5.25% to 8.25% chlorine.

Calcium Hypochlorite: This solid compound, commonly known as pool shock, is more concentrated than bleach and can be used to create a chlorine solution for water treatment.

Application:
Liquid Chlorine: Add a specified number of drops of chlorine bleach per gallon of water, then mix thoroughly and let stand for a designated contact time before consuming.

Calcium Hypochlorite: Dissolve a measured amount of calcium hypochlorite in water to create a chlorine solution, then add the solution to untreated water and let it sit for a specified time before drinking.

Advantages:
Effective against a broad spectrum of microorganisms.

Relatively inexpensive and readily available.

Easy to use and store, especially in the form of bleach.

Limitations:

Leaves a residual taste and odor in treated water.

Can react with organic matter to form disinfection by-products (DBPs) like trihalomethanes (THMs) in high concentrations.

Effectiveness can be affected by water temperature, pH, and turbidity.

2. Iodine Treatment:

Description: Iodine is a chemical disinfectant that kills a wide range of waterborne pathogens, including bacteria, viruses, and some protozoa. It is typically used in the form of iodine tablets or iodine tincture.

Application:

Iodine Tablets: Drop iodine tablets into a container of water, then wait for the specified contact time before drinking. The tablets release iodine into the water to disinfect it.

Iodine Tincture: Add a few drops of iodine tincture (2% iodine solution) per liter of water, mix well, and let stand for the recommended contact time before consumption.

Advantages:

Effective against a broad range of pathogens, including Giardia and Cryptosporidium.

Lightweight and portable, making it suitable for backpacking and emergency kits.

Does not leave a residual taste or odor in treated water.

Limitations:

Not recommended for long-term or continuous use due to potential health risks associated with high iodine intake, such as thyroid issues.

May cause allergic reactions in some individuals.

Pregnant women, individuals with thyroid disorders, and those with iodine sensitivity should avoid iodine treatment.

When using chemical treatment methods for water purification, it's essential to follow the manufacturer's instructions carefully, including dosage, contact time, and safety precautions. Additionally, it's advisable to filter turbid or cloudy water before chemical treatment to improve effectiveness. Rotate stored chemicals regularly and monitor expiration dates to ensure efficacy in emergency situations.

UV Sterilization and Solar Disinfection

UV sterilization and solar disinfection are effective water purification methods for preppers preparing for survival situations. Here's an in-depth look at each:

1. UV Sterilization:

Description: UV sterilization involves using ultraviolet light to destroy the DNA of microorganisms, rendering them unable to reproduce and cause infections. UV-C light with a wavelength of 254 nanometers is most effective for water disinfection.

Equipment:

UV Water Purifiers: These devices typically consist of a UV lamp enclosed in a quartz sleeve, a chamber for water flow, and sensors to monitor water quality and lamp functionality.

Application:

Water is passed through the UV chamber, where it is exposed to UV-C light for a specified period. The intensity and duration of UV exposure depend on factors such as flow rate, water quality, and lamp power.

UV sterilization is effective against bacteria, viruses, and protozoa, including Cryptosporidium and Giardia.

Advantages:

Provides rapid disinfection without the need for chemicals or heat.

Leaves no residual taste, odor, or chemical by-products in treated water.

Does not alter water chemistry or pH.

Low maintenance, with only periodic replacement of UV lamps required.

Limitations:

Requires electricity to operate, limiting its use in off-grid or remote areas unless powered by solar or battery.

Does not remove sediment or particulate matter from water, so pre-filtration may be necessary.

Effectiveness can be reduced by turbidity, color, and organic matter in water.

2. Solar Disinfection (SODIS):

Description: Solar disinfection, also known as SODIS, harnesses the power of sunlight to inactivate pathogens in water. The combined effects of UV radiation and heat from sunlight kill or deactivate microorganisms, making water safe for consumption.

Application:

Fill clear PET (polyethylene terephthalate) bottles with water from a contaminated source.

Place the bottles on a flat surface in direct sunlight for 6 hours on a sunny day or 2 consecutive days if the weather is cloudy.

The UV radiation and heat from the sun disinfect the water by killing bacteria, viruses, and parasites.

Advantages:

Simple and inexpensive method requiring only sunlight and PET bottles.

Suitable for off-grid and emergency situations where access to other purification methods may be limited.

No chemicals or specialized equipment required.

Limitations:

Requires clear skies and sufficient sunlight for effective disinfection, which may be challenging in certain climates or seasons.

Does not remove sediment, turbidity, or chemical contaminants from water.

Limited to treating small volumes of water and may not be practical for large-scale purification.

When utilizing UV sterilization or solar disinfection for water purification, it's important to ensure proper exposure to UV light and sunlight, respectively. Regular maintenance and monitoring of equipment are necessary to ensure consistent performance. Additionally, pre-filtration may be required to remove sediment and particulate matter before UV treatment or solar disinfection to optimize effectiveness.

Boiling Water Safely

Boiling water is one of the oldest and most reliable methods of purifying water, making it safe for consumption. For preppers preparing for survival situations, knowing how to boil water safely is essential. Here's a detailed guide on boiling water for purification:

1. Preparation:

  Selecting Water Source: Choose a water source that appears clear and free of visible contaminants. Avoid water sources located near industrial sites, agricultural fields, or areas with potential chemical contamination.

  Collection: Collect water in a clean container. If possible, use a container with a lid to minimize contamination during transportation.

  Filtration: If the water appears cloudy or contains sediment, filter it through a clean cloth or paper towel to remove particulate matter before boiling. This step helps improve the clarity and taste of the boiled water.

2. Boiling Process:

  Heat Source: Use a heat source such as a campfire, portable stove, or propane burner to heat the water. Ensure the heat source is stable and located in a well-ventilated area.

  Boiling Time: Bring the water to a rolling boil. A rolling boil is characterized by large bubbles continuously breaking the surface of the water.

Once boiling, maintain the heat and continue boiling the water for at least one minute.

Altitude Consideration: At higher altitudes, water boils at lower temperatures due to reduced atmospheric pressure. To compensate for this, extend the boiling time. For altitudes above 6,562 feet (2,000 meters), boil water for three minutes to ensure effective disinfection.

3. Cooling and Storage:

Cooling: Allow the boiled water to cool before transferring it to clean, covered containers for storage. Using plastic containers that may leach chemicals into the water should be avoided

Storage: Store the boiled water in a cool, dark place away from direct sunlight and potential sources of contamination. Use the water within 24 hours for best quality and safety.

4. Testing for Contaminants:

While boiling water effectively kills most pathogens, it may not remove chemical contaminants or heavy metals. If there are concerns about chemical contamination, consider using additional purification methods such as activated carbon filtration or chemical treatment.

In situations where water quality is questionable or if individuals experience symptoms of waterborne illness after drinking boiled water, seek alternative water sources or consult with local authorities for guidance.

5. Safety Precautions:

Exercise caution when handling boiling water to avoid burns or scalds.

Ensure children and pets are kept away from the boiling water and heat source to prevent accidents.

Use appropriate cookware and utensils designed for boiling water to minimize the risk of injury or contamination.

Boiling water is a reliable and effective method for purifying water in emergency situations. By following these guidelines and practicing proper water safety measures, preppers can ensure access to clean and safe drinking water during survival scenarios.

CHAPTER FOUR

Water Storage Strategies

Selecting and Preparing Water Containers

Selecting and preparing water containers is a critical aspect of water storage strategy for preppers preparing for survival scenarios. Proper containers ensure water remains clean, safe, and readily available for consumption. Here's an extensive guide on selecting and preparing water containers:

1. Selecting Water Containers:

Food-Grade Material: Choose containers made from food-grade materials such as high-density polyethylene (HDPE), polypropylene (PP), or stainless steel. Avoid containers made from materials that may leach harmful chemicals into the water, such as certain plastics or metals.

Size and Capacity: Select containers of appropriate size and capacity based on your water storage needs and available space. Consider factors such as the number of individuals in your group, estimated water consumption, and duration of survival scenarios.

Durability: Opt for durable containers that can withstand rough handling, temperature fluctuations, and prolonged storage. Look for containers with reinforced seams, thick walls, and impact-resistant construction to ensure longevity and reliability.

Portability: Choose containers that are easy to transport and maneuver, especially if you need to relocate or evacuate quickly. Consider collapsible or stackable containers for efficient storage and transportation.

Sealing Mechanism: Select containers with secure sealing mechanisms such as screw-on lids, snap-lock closures, or gaskets to prevent leakage and contamination. Ensure the lids are tight-fitting and provide a reliable seal to maintain water quality.

Transparency: Preferably choose transparent or translucent containers that allow you to monitor water levels and visually inspect the water for clarity, color, and signs of contamination without opening the container.

Accessibility: Consider accessibility features such as built-in handles, spigots, or pour spouts that facilitate easy dispensing of water without the need for additional equipment or tools.

2. Preparing Water Containers:

Cleaning and Sanitizing: Thoroughly clean and sanitize new or used water containers before use to remove any residues, contaminants, or bacteria. Wash containers with hot, soapy water, rinse thoroughly, and sanitize with a diluted bleach solution (1 teaspoon of bleach per gallon of water). Allow containers to air dry completely before filling with water.

Avoiding Contamination: Take precautions to prevent contamination of water containers during storage and handling. Keep containers sealed and stored in a clean, dry, and well-ventilated area away from direct sunlight, chemicals, pests, and other potential sources of contamination.

Labeling and Dating: Label each water container with the date of filling and any other relevant information such as water source, treatment method, or expiration date. Use waterproof and fade-resistant markers or labels to ensure legibility over time.

Rotating Stock: Implement a rotation schedule to ensure water stored in containers remains fresh and potable. Regularly inspect containers for signs of damage, deterioration, or leakage, and replace any compromised containers as needed. Use a "first in, first out" (FIFO) approach to consume and replenish water supplies to maintain freshness and quality.

By carefully selecting and preparing water containers, preppers can effectively store and preserve water for survival scenarios. Investing in high-quality, food-grade containers, and following proper cleaning and maintenance procedures ensure water remains safe, clean, and accessible when needed most.

Long-Term Water Storage Solutions

Long-term water storage solutions are essential for preppers preparing for extended survival scenarios where access to clean water may be limited or unavailable. Here's a detailed guide on long-term water storage strategies for preppers:

1. Selecting Suitable Containers:

Food-Grade Containers: Choose containers made from food-grade materials such as high-density polyethylene (HDPE), polypropylene (PP), or stainless steel. Avoid containers made from materials that may leach harmful chemicals into the water, such as certain plastics or metals.

Large Capacity: Opt for containers with large storage capacities to accommodate long-term water needs. Consider factors such as the number of individuals in your group, estimated water consumption, and duration of survival scenarios when selecting container size.

Durability: Prioritize durable containers that can withstand prolonged storage, temperature fluctuations, and rough handling. Look for containers with reinforced seams, thick walls, and impact-resistant construction to ensure longevity and reliability.

Sealing Mechanism: Select containers with secure sealing mechanisms such as screw-on lids, snap-lock closures, or gaskets to prevent leakage and contamination. Ensure the lids provide a tight seal to maintain water quality over time.

UV Protection: Choose containers that offer protection against UV radiation to prevent degradation of the container material and potential contamination of the stored water. Look for containers with UV-resistant coatings or opaque colors that block sunlight.

Accessibility: Consider accessibility features such as built-in handles, spigots, or pour spouts that facilitate easy dispensing of water without the need for additional equipment or tools.

2. **Water Treatment and Preservation**:
Water Quality: Start with clean, potable water from a reliable source whenever possible. If using untreated water, ensure it is filtered and treated to

remove contaminants and pathogens before storage.

Water Treatment: Treat water with chlorine bleach, iodine tablets, or other water purification methods to disinfect and preserve it for long-term storage. Follow manufacturer instructions for proper dosage and treatment duration.

Preservation Methods: Consider adding stabilizers such as hydrogen peroxide or sodium hypochlorite to prevent the growth of algae, bacteria, and other microorganisms that can degrade water quality over time. Rotate stock periodically to maintain freshness and potability.

3. Storage Conditions:
Location: Store water containers in a cool, dark, and well-ventilated area away from direct sunlight, chemicals, pests, and other potential sources of contamination. Choose a location with stable temperatures to minimize fluctuations that can affect water quality.

Elevation: Elevate water containers off the ground to prevent contact with moisture, pests, and potential contaminants. Use pallets, shelves, or other raised platforms to elevate containers and facilitate airflow around them.

Insulation: Insulate water storage areas to regulate temperatures and prevent freezing in cold climates or excessive heat buildup in warm climates. Use insulation materials such as foam panels, blankets, or reflective barriers to maintain optimal storage conditions.

4. Maintenance and Monitoring:

Regular Inspection: Periodically inspect water containers for signs of damage, deterioration, or leakage. Check seals, lids, and fittings for tightness and integrity, and replace any compromised containers or components as needed.

Rotation Schedule: Implement a rotation schedule to ensure water stored in containers remains fresh and potable. Use a "first in, first out" (FIFO) approach to consume and replenish water supplies to maintain freshness and quality.

Water Testing: Test stored water periodically for purity and potability using water testing kits or laboratory analysis. Monitor water quality indicators such as taste, odor, color, and clarity, and take corrective actions if abnormalities are detected.

By implementing these long-term water storage strategies, preppers can effectively store and preserve water for extended survival scenarios. Prioritizing proper container selection, water treatment, storage conditions, and maintenance practices ensures water remains clean, safe, and accessible when needed most.

Rotating Water Supplies

Rotating water supplies is a crucial aspect of water storage strategy for preppers preparing for survival scenarios. Proper rotation ensures that stored water remains fresh, clean, and safe for consumption over time. Here's a detailed guide on rotating water supplies:

1. Why Rotate Water Supplies:
   Maintain Freshness: Water stored for extended periods can become stale or develop off-flavors due to exposure to air and environmental contaminants. Rotating water supplies helps maintain freshness and taste.

   Prevent Contamination: Stored water is susceptible to contamination from microorganisms, algae, and other pollutants over time. Regular rotation minimizes the risk of contamination and ensures water remains safe for consumption.

   Ensure Potability: Water quality may deteriorate over time due to factors such as temperature fluctuations, exposure to light, and bacterial growth. Rotating water supplies helps ensure stored water remains potable and free from harmful pathogens.

   Optimize Shelf Life: Proper rotation extends the shelf life of stored water by preventing degradation and spoilage. Fresh water is essential for hydration, cooking, and hygiene in survival scenarios, making regular rotation essential for preparedness.

2. Rotation Schedule:

Frequency: Establish a rotation schedule based on the shelf life of stored water and your anticipated water consumption. Aim to rotate water supplies every six months to one year, depending on storage conditions and container integrity.

First In, First Out (FIFO): Use a FIFO approach to rotate water supplies, consuming the oldest water first before accessing newer supplies. This ensures that water is used and replenished in a timely manner to maintain freshness.

Calendar Reminders: Set calendar reminders or schedule rotation dates to ensure regular and systematic rotation of water supplies. Mark containers with the date of filling to track storage duration and prioritize consumption accordingly.

Seasonal Considerations: Consider rotating water supplies seasonally to coincide with changes in temperature and environmental conditions. Higher temperatures and humidity levels can accelerate water degradation and bacterial growth, necessitating more frequent rotation during warmer months.

3. Rotation Process:

Inspect Containers: Before rotating water supplies, inspect containers for signs of damage, deterioration, or contamination. Check seals, lids, and fittings for tightness and integrity, and replace any compromised containers or components as needed.

Empty and Clean: Empty water containers completely and clean them thoroughly with hot, soapy water to remove any residues or impurities. Rinse containers with clean water and sanitize them with a diluted bleach solution (1 teaspoon of bleach per gallon of water). Allow containers to air dry completely before refilling.

Refill with Fresh Water: Refill cleaned and sanitized containers with fresh, potable water from a reliable source. Treat the water with chlorine bleach or other water purification methods if necessary to ensure safety and cleanliness.

Label and Date: Label each container with the date of filling and any other relevant information such as water source, treatment method, or expiration date. Use waterproof and fade-resistant markers or labels to ensure legibility over time.

Store Properly: Store rotated water containers in a cool, dark, and well-ventilated area away from direct sunlight, chemicals, pests, and other potential sources of contamination. Choose a location with stable temperatures to minimize fluctuations that can affect water quality.

4. Monitoring and Maintenance:

Regular Inspection: Periodically inspect rotated water supplies for signs of degradation, contamination, or leakage. Monitor water quality indicators such as taste, odor, color, and clarity, and take corrective actions if abnormalities are detected.

Replenish as Needed: Replenish water supplies as needed to maintain adequate reserves for survival scenarios. Consider factors such as changes in group size, water consumption rates, and anticipated duration of survival situations when replenishing supplies.

Water Testing: Test rotated water periodically for purity and potability using water testing kits or laboratory analysis. Conduct tests for bacteria, parasites, and chemical contaminants to ensure water remains safe for consumption.

By implementing a systematic approach to rotating water supplies, preppers can ensure that stored water remains fresh, clean, and safe for consumption in survival scenarios. Prioritize regular inspection, cleaning, and replenishment to maintain water quality and optimize preparedness for emergencies.

Creative Storage Options for Limited Spaces

In survival scenarios where space is limited, preppers must get creative with water storage solutions to maximize storage capacity while ensuring accessibility and safety. Here are several creative storage options for limited spaces:

1. Stackable Water Containers:

Opt for stackable water containers designed to maximize vertical space utilization. These containers are typically rectangular or square-shaped and feature interlocking lids or grooves that allow them to be stacked securely on top of each other without tipping or collapsing.

2. Under-Bed Storage:

Utilize the space under beds for storing water containers. Choose low-profile containers that fit snugly beneath the bed frame, such as flat or shallow storage bins or collapsible water bladders. Ensure containers are properly sealed and protected from dust, pests, and other contaminants.

3. Overhead Storage

Install overhead shelving or racks in closets, garages, or other areas with high ceilings to store water containers. Use heavy-duty hooks, brackets, or straps to secure containers to the overhead storage system and prevent them from falling or shifting.

4. Behind Furniture or Appliances:

Place water containers behind large furniture pieces or appliances such as sofas, dressers, refrigerators, or washing machines. This hidden storage solution makes use of otherwise unused space and keeps water containers out of sight while maintaining accessibility.

5. Corner Storage:

Maximize corner spaces in rooms or storage areas by placing triangular or wedge-shaped water containers. These specially designed containers fit snugly into corners, optimizing space efficiency and allowing easy access to stored water.

6. Vertical Storage Racks:

Install vertical storage racks or shelves specifically designed for storing water containers. These racks typically feature slots or compartments sized to accommodate standard water jugs or bottles, allowing you to organize and store containers vertically to save floor space.

7. Roll-Out Storage Carts:

Invest in roll-out storage carts or drawers that can be installed under countertops or inside cabinets. These carts provide convenient access to water containers while keeping them organized and out of the way when not in use. Choose carts with sturdy construction and smooth-gliding mechanisms for durability and ease of use.

8. DIY PVC Pipe Storage System:

Build a custom storage system using PVC pipes to create vertical slots for holding water bottles or jugs. Cut PVC pipes into sections of suitable length, then secure them vertically to a wall or frame using brackets or adhesive mounts. This DIY solution allows you to customize the size and configuration of the storage system to fit your space requirements.

9. Hanging Storage Bags or Pouch

Hang water storage bags or pouches from hooks or racks mounted on walls or ceilings. These flexible and lightweight containers can be filled with water and suspended in the air to save floor space. Ensure proper support and reinforcement to prevent overloading and potential damage.

10. Dual-Purpose Furniture:

Choose furniture pieces that double as water storage containers, such as ottomans, benches, or coffee tables with hidden compartments. These multifunctional pieces of furniture provide concealed storage for water while serving practical purposes in living spaces.

By implementing these creative storage options for limited spaces, preppers can effectively store and access water supplies while optimizing available space in survival scenarios. Prioritize proper container selection, organization, and maintenance to ensure water remains clean, safe, and readily available for consumption when needed most.

CHAPTER FIVE

Water Conservation

Importance of Water Conservation in Survival Scenarios

Water conservation is crucial in survival scenarios for preppers to ensure the availability of clean and safe drinking water for an extended period. Here's a comprehensive guide outlining the importance of water conservation and practical tips for preppers:

1. Limited Water Supply:
During survival situations, getting access to clean water may be limited or non-existent. Conserving water helps stretch available supplies and ensures a steady source of hydration for drinking, cooking, and hygiene.

2. Extended Survival Duration
By conserving water, preppers can prolong their survival duration, increasing their chances of staying hydrated and maintaining overall health until rescue or the situation improves.

3. Preservation of Health:
Clean water is essential for maintaining health and preventing waterborne illnesses. Conserving water reduces the risk of dehydration, which can

lead to fatigue, weakness, and other health complications.

4. Environmental Considerations:

Water conservation in survival scenarios helps minimize environmental impact by reducing the need to extract water from natural sources. Conserving water also preserves ecosystems and wildlife habitats.

5. Energy Efficiency:

Many water purification methods require energy, whether it's boiling water over a fire or operating filtration devices. By conserving water, preppers reduce the energy demands associated with water treatment, contributing to energy efficiency.

6. Emergency Preparedness:

Practicing water conservation in everyday life prepares preppers for emergency situations where water resources are scarce or compromised. Developing habits of water conservation ensures readiness for unexpected events.

7. Sustainable Living

Water conservation aligns with principles of sustainable living by promoting responsible use of resources. Preppers who prioritize water conservation contribute to environmental stewardship and resilience in the face of challenges.

Practical Tips for Water Conservation in Survival Scenarios:

1. Rationing:
Establish daily water rations based on individual needs and available supply. Consume water conservatively and avoid wasteful practices.

2. Reuse and Recycling:
Collect and reuse water whenever possible. For example, capture rainwater for drinking, cooking, and hygiene purposes. Repurpose wastewater from activities like washing dishes for flushing toilets or watering plants.

3. Personal Hygiene Practices:
Practice water-efficient hygiene habits, such as taking short showers or using wet wipes for cleaning. Use minimal water for washing dishes and laundry, and consider air-drying items to conserve water.

4. Leak Prevention:
Check for and repair leaks in water storage containers, plumbing fixtures, and irrigation systems to prevent water loss. Every drop saved contributes to overall water conservation efforts.

5. Landscaping and Gardening:
Prioritize drought-resistant plants and water-efficient landscaping techniques in survival gardens. Mulching, drip irrigation, and strategic

planting help conserve water while supporting food production.

6. Education and Training

Educate yourself and your group members on the importance of water conservation and train in water-saving techniques. Share knowledge and best practices to ensure everyone contributes to conservation efforts.

7. Innovation and Adaptation

Explore innovative solutions for water conservation, such as alternative water sources like dew collection or condensation traps. Adapt to changing circumstances and conditions to optimize water use efficiency.

By prioritizing water conservation in survival scenarios, preppers can maximize the sustainability of their resources, increase their resilience to adversity, and maintain essential hydration and hygiene practices for long-term survival.

Strategies for Reducing Water Usage

Reducing water usage is essential for preppers in survival scenarios to conserve precious water resources and ensure long-term sustainability. Here are various strategies for reducing water usage:

1. Rationing and Planning:

   Establish a daily water ration based on the number of people in your group and the available water supply. Plan ahead to ensure everyone has access to enough water for drinking, cooking, and hygiene needs.

2. Efficient Personal Hygiene:

   Take shorter showers or sponge baths to minimize water usage. Turn off the water while lathering or shampooing, and use a low-flow showerhead if available. Consider using wet wipes or dry shampoo as alternatives for bathing.

3. Water-Saving Appliances and Fixtures

   Install water-saving devices such as low-flow faucets, showerheads, and toilets to reduce water consumption. These fixtures limit the flow of water without compromising performance, helping conserve water in both everyday life and survival scenarios.

4. Greywater Recycling:

   Collect and reuse greywater from activities like washing dishes, laundry, and bathing for non-potable purposes such as flushing toilets or watering plants. Implementing greywater recycling systems helps maximize water efficiency and minimizes waste.

5. Leak Detection and Repair:
   Regularly inspect water storage containers, plumbing fixtures, and irrigation systems for leaks. Repair any leaks promptly to prevent water loss and ensure efficient water usage. Even small leaks can lead to significant water waste over time.

6. Rainwater Harvesting
   Rainwater is collected from rooftops or other surfaces through the use of rain barrels or cisterns. Rainwater harvesting provides a supplementary water source for non-potable uses like gardening, cleaning, and flushing toilets, reducing reliance on treated water for these purposes.

7. Xeriscaping and Water-Efficient Landscaping:
   Design gardens and landscapes with drought-resistant plants native to your region. Implement xeriscaping techniques such as mulching, soil amendment, and efficient irrigation systems to minimize water requirements for landscaping while maintaining aesthetic appeal.

8. Water-Efficient Cooking and Cleaning Practices:
   Use minimal water when cooking by using tight-fitting lids, steaming instead of boiling, and reusing cooking water when possible. When washing dishes, scrape food scraps off plates before washing, and use a basin or plug in the sink to conserve water.

9. Educational Outreach and Training:
   Educate group members or community members on the importance of water conservation and train them in water-saving techniques. Encourage everyone to participate in conservation efforts and share tips and best practices for reducing water usage.

10. Innovative Solutions and Adaptation:
   Explore innovative water-saving solutions tailored to your specific survival scenario. Consider alternative water sources like dew collection, condensation traps, or water purification techniques that minimize water loss and maximize efficiency. By implementing these strategies for reducing water usage, preppers can effectively conserve water resources, increase resilience in survival scenarios, and ensure sustainable water management for long-term survival.

Repurposing Water for Multiple Uses

Repurposing water for multiple uses is a crucial aspect of water conservation for preppers in survival scenarios. By maximizing the efficiency of water usage, preppers can stretch their water supply and ensure sustainability for longer periods.

1. Greywater Recycling

   Definition: Greywater refers to wastewater generated from activities like bathing, laundry, and dishwashing that does not contain significant amounts of human waste.

Collection: Capture greywater from sinks, showers, and washing machines using a separate drainage system or by manually collecting it in containers.

Treatment: Greywater can be filtered or treated using simple methods such as settling or filtration to remove debris and impurities before reuse.

Reuse: Repurpose treated greywater for non-potable purposes such as flushing toilets, watering plants, and cleaning outdoor surfaces.

2. Rainwater Harvesting:

Collection: Collect rainwater runoff from rooftops, gutters, and other surfaces using rain barrels, cisterns, or underground tanks.

Filtration: Filter collected rainwater to remove sediment, debris, and contaminants using screens, mesh filters, or commercial filtration systems.

Storage Store filtered rainwater in clean, covered containers for later use in drinking, cooking, and hygiene, or for non-potable purposes like irrigation and landscaping.

3. Water from Cooking and Cleaning

Cooking Water: After cooking food such as pasta or vegetables, allow the cooking water to cool and reuse it for tasks like watering plants or adding moisture to compost piles.

Cleaning Water: Water used for cleaning dishes or surfaces can be reused for secondary cleaning tasks, such as washing outdoor equipment or rinsing off gardening tools.

4. Personal Hygiene Practices:

Bathing: Use a basin or tub to collect water while bathing, then reuse it for flushing toilets, cleaning, or watering plants.

Laundry: Reuse laundry, rinse water for subsequent loads or for cleaning outdoor items like vehicles, tools, or patio furniture.

5. Animal Watering:

Use water collected from various sources, such as rainwater or greywater, to provide hydration for pets and livestock. Ensure the water is filtered and free from contaminants before offering it to animals.

6. Educational Outreach and Training:

Educate group members or community members on the importance of repurposing water for multiple uses and train them in water-saving techniques. Encourage everyone to participate in conservation efforts and share tips and best practices for maximizing water efficiency.

7. Monitoring and Maintenance:

Regularly monitor water collection, storage, and distribution systems to ensure proper functioning and prevent contamination. Clean and maintain equipment such as rain barrels, filters, and storage tanks to optimize performance and longevity.

By repurposing water for multiple uses, preppers can significantly reduce water waste, conserve precious resources, and enhance their resilience in survival scenarios. Implementing these practices requires careful planning, resourcefulness, and a commitment to sustainable water management.

Hygiene and Personal Care Tips

In survival scenarios where water resources are limited, maintaining hygiene and personal care becomes essential for preppers to prevent illness and ensure overall well-being. Here are detailed hygiene and personal care tips tailored for water conservation:

1. Prioritize Essential Hygiene Practices:
   Hand Hygiene: Wash hands thoroughly with soap and water before handling food, eating, or after using the restroom. Use hand sanitizer with at least 60% alcohol if water is scarce.
   Oral Hygiene: Brush teeth with minimal water usage by wetting the toothbrush, brushing, and then using a small amount of water to rinse. Consider using mouthwash as an alternative to conserve water.
   Body Hygiene: Practice sponge baths or use wet wipes to clean the body, focusing on areas prone to sweat and odor. Use minimal water and soap, and prioritize areas such as the face, underarms, and groin.

2. **Optimize Showering Techniques**:

Shorter Showers: Take quick showers to minimize water usage. Wet the body, turn off the water while applying soap, shampoo, or conditioner, then rinse quickly.

Low-Flow Showerheads: Install low-flow showerheads to reduce water consumption without sacrificing water pressure. These fixtures help conserve water while still providing an effective showering experience.

Navy Showers: Adopt the navy shower technique, which involves wetting the body, turning off the water while soaping up, then quickly rinsing off. This method saves significant amounts of water compared to traditional showering.

3. **Water-Efficient Cleaning Practices:**

Dishwashing: Use minimal water for dishwashing by filling a basin with soapy water for washing and a separate basin with clean water for rinsing. Alternatively, use disposable plates and utensils to conserve water.

Laundry : Wash clothing only when necessary and use a minimal amount of water. Consider hand-washing small loads using a basin or sink, and air dry clothing whenever possible to reduce energy consumption.

Surface Cleaning: Use wet wipes, spray bottles with diluted cleaning solutions, or reusable cleaning cloths for surface cleaning. Focus on high-touch areas and areas prone to dirt and germs to maintain cleanliness.

4. Waterless Hygiene Alternatives:

Dry Shampoo: Use dry shampoo to refresh hair between washes and reduce the need for frequent hair washing. Dry shampoo absorbs excess oil and provides a clean appearance without water.

Wet Wipes: Keep a supply of wet wipes or baby wipes for quick and convenient personal hygiene. Use wipes for cleaning hands, face, and body when water is not readily available.

5. Community Hygiene Practice

Shared Facilities: Coordinate hygiene practices among group members to optimize water usage. Share resources such as hand sanitizer, wet wipes, and cleaning supplies to conserve water collectively.

Education and Training: Educate group members on the importance of water conservation for personal hygiene and train them in water-saving techniques. Encourage everyone to contribute to conservation efforts and share tips for maximizing water efficiency.

By implementing these hygiene and personal care tips, preppers can maintain cleanliness and well-being while conserving water resources in survival scenarios. Prioritizing essential hygiene practices and adopting water-efficient techniques are essential for optimizing water usage and ensuring sustainability in challenging environments.

CHAPTER SIX

Emergency Water Procurement

In emergency situations, securing a safe and reliable water supply is essential for survival. Emergency water procurement involves the process of finding, collecting, treating, and storing water when traditional water sources are unavailable or compromised.

Short-Term Solutions for Water Acquisition

Friendly guide to short-term solutions for water acquisition in a survival scenario, tailored for beginners:

Short-Term Solutions for Water Acquisition
In a survival situation, finding safe water quickly is essential for your well-being. Here are some simple and effective short-term solutions to help you acquire water:

1. Collect Rainwater:
   Use clean containers or improvised rain catchment systems to collect rainwater during showers or storms.
   Rainwater is generally safe to drink if collected in a clean container, away from contaminants.

2. Find Natural Water Sources:

Look for rivers, streams, or lakes nearby. Moving water (like from a river) is often safer than stagnant water.

Always purify water from natural sources before drinking to remove bacteria and other contaminants.

3. Use Water Purification Tablets or Drops:

Carry water purification tablets or drops in your survival kit.

Follow the instructions to treat water from suspect sources to make it safe for drinking.

4. Boil Water:

Boiling water is an effective way to kill harmful bacteria, viruses, and parasites.

Bring water to a rolling boil for at least one minute (or longer at higher altitudes) to ensure it's safe.

5. Build a Solar Still:

Create a solar still to extract water from moist soil or vegetation.

Dig a hole, place a container in the center, cover the hole with plastic wrap, and secure the edges. Condensation will collect inside the container.

6. Use Filtration Systems:

Use portable water filters or purifiers designed for outdoor use.

These devices can remove bacteria, protozoa, and other contaminants from water.

7. Tap Trees for Water:

Certain trees, like birch or maple, can produce sap that is safe to drink.

Use a knife to tap into the tree trunk and collect the sap in a container.

8. Conserve Water:

Minimize water loss by using water sparingly for drinking and essential hygiene.

Avoid activities that can lead to excessive sweating to conserve body fluids.

9. Use Household Items for Collection:

Use clean clothing, tarps, or containers to gather dew or moisture overnight.

Squeeze the collected moisture into a container for drinking.

10. Stay Alert for Signs of Water:

Look for signs of wildlife congregating or vegetation growing near water sources.

Listen for the sound of running water or follow game trails that may lead to water sources.

Remember, while these short-term solutions can help in emergencies, it's crucial to prioritize water purification and ensure the water you consume is safe. Always be prepared with proper gear and knowledge to acquire and treat water for survival needs.

Improvising Water Collection and Filtration Devices

In a survival situation, being able to improvise water collection and filtration devices using readily available materials can be essential for obtaining safe drinking water. Here's a detailed guide on how to improvise these devices for preppers:

Water Collection Devices:
1. Rainwater Collection:
   Materials Needed: Large clean containers (e.g., buckets, barrels), plastic sheeting, rope or cord.
   Method:
   Place a clean container directly in the path of rainwater to collect it.
   For larger-scale collection, create a rain catchment system using plastic sheeting.
   Secure the plastic sheeting over a frame or between two elevated points, allowing rainwater to run off into a container placed below.
   Use rope or cord to anchor and secure the plastic sheeting in place.

2. Solar Still:
   Materials Needed: Digging tools (shovel or trowel), clear plastic sheeting, rocks or weights.
   Method:
   Dig a hole in moist soil or near a water source.

Place a clean container in the center of the hole.

Cover the hole with clear plastic sheeting, ensuring it's tightly sealed around the edges.

Place rocks or weights on the edges of the sheeting to create a slope towards the container.

Condensation will form on the underside of the plastic and drip into the container, providing distilled water.

3. Transpiration Bag (Tree Bag):

Materials Needed: Plastic bag or large leafy branch, cord or vine.

Method:

Secure a plastic bag tightly around a leafy branch of a tree.

Tie the opening of the bag with a cord or vine to create a seal.

Sunlight will cause moisture to evaporate from the leaves, condensing inside the bag.

Water droplets will collect at the bottom of the bag and can be collected for drinking.

Water Filtration Devices:
1. Improvised Charcoal Filter:

Materials Needed: Charcoal (from a campfire), clean cloth or bandana, container.

Method:

Crush charcoal into small pieces and place a layer of it inside a container.

Place a clean cloth or bandana over the charcoal layer.

Pour water slowly through the cloth and charcoal filter.

The charcoal will help absorb impurities and improve water quality.

2. Sand and Gravel Filter:

Materials Needed: Clean sand, gravel or small rocks, empty plastic bottles or containers.

Method:

Cut the bottom off a plastic bottle to create a funnel.

Layer clean sand, gravel, and small rocks inside the bottle, starting with the coarsest material at the bottom.

Pour contaminated water slowly through the filter.

The sand and gravel will trap sediment and larger particles, improving water clarity.

3. Boiling Method:

Materials Needed: Firewood, metal container or pot.

Method:

Build a fire and allow it to burn down to embers.

Place a metal container or pot filled with water over the fire.

Bring the water to a rolling boil and continue boiling for at least 1 minute (longer at higher altitudes).

Allow the water to cool before drinking.

Additional Tips:
 Always treat collected water (even from natural sources) before drinking to eliminate harmful bacteria, viruses, and parasites.
 Use improvised water collection and filtration methods as a temporary solution until you can access more reliable water sources or purification tools.
 Regularly inspect and maintain improvised devices to ensure they are clean and effective at providing safe drinking water.

By improvising water collection and filtration devices using basic materials and techniques, preppers can enhance their ability to procure safe drinking water in emergency situations. Practice these methods before an actual survival scenario to build confidence and readiness.

Wilderness Survival Techniques for Finding Water

When you're in a wilderness survival situation, finding water is a top priority for ensuring hydration and survival. Here are detailed techniques and tips for locating water sources in the wilderness:
1. Conducting a Water Search:
  Stay Calm and Assess: Keep calm and assess your surroundings for potential water sources such as rivers, streams, lakes, and wetlands.

Follow Wildlife: Look for signs of wildlife or animal tracks that may lead you to water sources, as animals need water to survive.

Listen for Sounds: Listen for the sound of running water, which can indicate the presence of nearby streams or rivers.

2. Natural Water Sources:

Rivers and Streams: Follow downstream to locate flowing water, which is generally safer to drink than stagnant water.

Lakes and Ponds: Look for bodies of water such as lakes or ponds, but avoid stagnant or visibly polluted water.

Springs and Seeps: Check for springs or seeps where groundwater naturally reaches the surface, often found in valleys or low-lying areas.

3. Signs of Water in the Wilderness:

Vegetation: Look for lush green vegetation or trees that require abundant water, as they may indicate the presence of nearby water sources.

Animal Activity: Watch for birds, insects, and other wildlife congregating near water sources, as they can lead you to water.

Damp Ground: Search for damp or moist soil, especially in shaded areas, which may indicate underground water.

4. Collecting Rainwater:

Rain Catchment: Use tarps, ponchos, or other waterproof materials to collect rainwater during storms or showers.

Containers: Use clean containers or improvised vessels to collect rainwater from natural depressions or rock formations.

5. Solar Still:

Digging a Solar Still: Create a solar still by digging a hole in moist soil, placing a container in the center, covering it with plastic, and collecting condensation.

Transpiration Bag: Secure a plastic bag tightly around a leafy branch to collect moisture evaporating from the leaves.

6. Water Sources in Plants:

Edible Plants: Some plants, like cacti or vines, may contain water that can be extracted by squeezing or cutting the plant.

Tree Sap: Tap into certain tree species to collect sap, which is usually safe to drink.

7. Filtration and Purification:

Improvised Filters: Use natural materials like sand, gravel, and charcoal to create improvised filters for removing sediment and impurities from collected water.

Boiling: Boil water over a fire to kill bacteria, viruses, and parasites. Let the water cool before drinking.

8. Avoiding Contaminated Water:

Polluted Water: Avoid drinking water that appears stagnant, murky, or polluted with visible debris or contaminants.

Animal Waste: Stay away from water sources contaminated with animal waste or carcasses.

9. Conserving Water:
Hydration: Conserve body moisture by avoiding excessive physical activity and minimizing sweat production.
Water Storage: Store collected water in clean containers or natural depressions to prevent evaporation and contamination.

10. Use Caution and Test Water:
Safety First: Always treat collected water (even from natural sources) before drinking to ensure it's safe.
Testing: Use water purification tablets, portable filters, or chemical treatments to purify and test water for safety.

By employing these wilderness survival techniques for finding water, preppers can increase their chances of locating safe drinking water in emergency situations. Practice these methods beforehand to build confidence and readiness for wilderness survival scenarios.

CHAPTER SEVEN

Maintaining Water Quality

In survival scenarios and emergency situations, access to clean and safe drinking water is essential for sustaining life and ensuring good health. As a prepper, understanding how to maintain water quality is paramount to surviving and thriving in challenging environments. This guide explores the importance of water quality maintenance and provides essential tips and strategies for preppers.

Why Water Quality Matters:
Water is a fundamental resource required for survival, playing a critical role in various bodily functions, from hydration to regulating body temperature and supporting organ functions. However, not all water sources are safe for consumption. Contaminated water can harbor harmful pathogens, chemicals, and pollutants that pose significant health risks when ingested.

Regular Inspection and Maintenance of Water Storage

Properly storing water is crucial for maintaining water quality and ensuring a safe and reliable water supply during emergency situations. Regular inspection and maintenance of water storage containers are essential to prevent contamination and preserve water quality. This guide outlines detailed steps for inspecting and maintaining water storage for preppers:

1. Inspection of Water Storage Containers:
a. Visual Inspection:

Regularly inspect the exterior of water storage containers for signs of damage, cracks, or leaks.

Check for algae growth, sediment buildup, or discoloration on the interior surfaces of the containers.

b. Lid and Closure Inspection:

Ensure that storage container lids have tight-fitting seals to prevent dust, debris, or insects from entering.

Inspect lid closures, gaskets, or seals for wear and tear, and replace them if necessary to maintain a secure closure.

c. Sanitation and Cleaning:

Clean water storage containers with soap and water before refilling them.

Use a mild bleach solution (1 teaspoon of unscented bleach per gallon of water) to disinfect containers and kill bacteria.

Rinse containers thoroughly with clean water to remove any residual bleach before refilling.

2. Maintenance of Water Storage Containers:
a. Rotation of Water Supply:
Regularly rotate stored water to prevent stagnation and ensure freshness.

Use and replace water stored for drinking purposes within recommended timeframes (typically every 6-12 months).

b. Temperature Control:
Store water containers in a cool, dark location away from direct sunlight to prevent bacterial growth and algae formation.

Avoid storing water containers in areas prone to extreme temperatures that could affect water quality.

c. Pest Control:
Keep water storage areas clean and free of debris, food crumbs, or other attractants that may attract pests.

Use pest control measures such as traps or repellents to deter insects or rodents from accessing water storage containers.

3. Regular Water Quality Testing:
a. Visual Inspection:
Visually inspect stored water for clarity, color, and odor. Cloudy or foul-smelling water may indicate contamination.

b. Water Testing Kits:

Use portable water testing kits to check for microbial contamination (e.g., bacteria, parasites) and chemical pollutants (e.g., chlorine, lead).

Follow manufacturer instructions to perform water tests and interpret results accurately.

4. Emergency Preparedness and Documentation:
a. Emergency Plan:

Include water storage inspection and maintenance protocols in your emergency preparedness plan.

Establish procedures for responding to water quality issues and implementing corrective actions.

b. Documentation:

Maintain records of water storage inspections, maintenance activities, and water quality test results.

Document dates of water rotation, cleaning procedures, and any observations related to water storage conditions.

By implementing regular inspection and maintenance practices for water storage, preppers can ensure a safe and reliable water supply during emergencies. Proactive measures to monitor and preserve water quality contribute to overall readiness and resilience in survival scenarios. Regularly assess and address potential issues to optimize water storage for long-term survival and well-being.

Testing Water Quality and Purity

Testing water quality and ensuring its purity is critical for preppers relying on stored or natural water sources during survival situations. Water testing helps identify potential contaminants and ensures safe drinking water for hydration and health. Here's a comprehensive guide on testing water quality and purity for preppers:

1. Importance of Water Testing:

Water testing is essential for detecting harmful contaminants, pathogens, and pollutants that may be present in water sources.

Testing helps assess water safety and determines the need for treatment or purification methods to make water potable.

2. Types of Water Testing:
a. Microbiological Testing:
Purpose: Detects bacteria, viruses, and parasites in water that can cause waterborne diseases.
Methods: Use portable water testing kits or laboratory analysis to check for coliform bacteria, E. coli, and other microbial contaminants.

b. Chemical Testing:
Purpose: Identifies chemical pollutants and toxic substances in water that may be harmful to human health.
Methods: Conduct tests for heavy metals (e.g., lead, arsenic), pesticides, herbicides, industrial chemicals, and disinfection byproducts (e.g., chlorine, chloramine).

c. Physical Testing:
Purpose: Assesses water clarity, color, odor, and taste to determine aesthetic quality.
Methods: Perform visual inspections and use sensory evaluation to detect abnormalities indicative of water contamination.

3. Water Testing Methods for Preppers:
a. Portable Water Testing Kits:
Features: Compact and easy-to-use kits designed for field testing of water quality parameters.
Tests Available: Microbial (bacteria and viruses), chemical (pH, chlorine levels), and physical (turbidity) parameters.
b. Laboratory Analysis:
Procedure: Collect water samples and send them to certified laboratories for comprehensive analysis.
Advantages: Provides accurate and detailed results for a wide range of contaminants and water quality parameters.

4. Steps for Conducting Water Quality Testing:

a. Sampling:
Collect water samples from the water source or storage container using clean, sterile containers.

Avoid contaminating the samples during collection to ensure accurate test results.

b. Testing Process:
Follow manufacturer instructions for using portable water testing kits or preparing water samples for laboratory analysis.

Conduct tests for specific contaminants based on the intended use (drinking water, cooking, hygiene).

c. Interpretation of Results:
Analyze test results to determine water quality and identify potential health risks associated with detected contaminants.

Compare results against recommended water quality standards (e.g., EPA guidelines) to assess compliance and safety.

5. Responding to Water Quality Issues:

a. Treatment and Purification:
Implement appropriate water treatment methods based on test results (e.g., boiling, filtration, chemical disinfection).

Use effective purification techniques to remove or neutralize contaminants and ensure safe drinking water.

b. Monitoring and Follow-Up:
Regularly monitor water quality and conduct follow-up testing to verify the effectiveness of treatment measures.

Adjust water treatment protocols as needed to maintain water purity and safety over time.

6. Documentation and Preparedness:
a. Record Keeping:

Maintain detailed records of water testing activities, including dates, test results, and corrective actions taken.

Use documentation to track water quality trends, identify recurring issues, and inform future preparedness efforts.

b. Emergency Planning:

Incorporate water quality testing protocols into emergency preparedness plans to ensure readiness for survival scenarios.

Establish procedures for responding to water quality emergencies and managing water resources effectively.

By implementing comprehensive water testing protocols, preppers can enhance their ability to assess and maintain water quality for sustained survival and well-being. Regular testing and proactive measures contribute to overall readiness and resilience in challenging environments. Prioritize water quality testing as a fundamental aspect of emergency preparedness and survival planning for preppers.

Treating Waterborne Illnesses

In survival scenarios or emergency situations, exposure to contaminated water can lead to waterborne illnesses caused by bacteria, viruses, parasites, and other pathogens. Knowing how to recognize and treat waterborne illnesses is essential for preppers relying on natural or stored water sources. This guide provides detailed information on identifying and treating common waterborne illnesses:

1. Common Waterborne Illnesses:
a. Bacterial Infections:
  Examples:
   Cholera: Causes severe diarrhea, vomiting, and dehydration.
   E. coli Infection: Can lead to gastrointestinal symptoms such as abdominal pain, diarrhea, and fever.
   Salmonella: Symptoms include diarrhea, fever, and abdominal cramps.
b. Viral Infections:
  Examples:
   Norovirus: Causes acute gastroenteritis with symptoms like diarrhea, nausea, and vomiting.
   Hepatitis A: Affects the liver and causes symptoms such as jaundice, fatigue, and abdominal pain.

c. Protozoan Parasites:
Examples:

Giardia lamblia: Causes giardiasis, characterized by diarrhea, abdominal cramps, and bloating.

Cryptosporidium: Leads to cryptosporidiosis with symptoms of watery diarrhea, nausea, and stomach cramps.

2. Symptoms of Waterborne Illnesses:
Diarrhea (often watery or bloody)
Nausea and vomiting
Abdominal cramps and pain
Fever and chills
Dehydration (dry mouth, reduced urine output, dizziness)

3. Treatment of Waterborne Illnesses:
a. Rehydration:
Encourage fluid intake to prevent dehydration and restore electrolyte balance.

Offer oral rehydration solutions (ORS) or electrolyte drinks to replenish lost fluids and minerals.

b. Antibiotic Treatment (if applicable):
Some bacterial infections may require antibiotic treatment prescribed by a healthcare provider.

Follow medical advice and complete the full course of antibiotics as directed.

c. Symptomatic Relief:
Use over-the-counter medications to manage symptoms such as diarrhea, nausea, and fever.

Rest and avoid strenuous activities to facilitate recovery.

4. Prevention of Waterborne Illnesses:
a. Water Treatment:
Purify and treat water from natural sources using methods like boiling, filtration, chemical disinfection, or UV sterilization.

Ensure stored water is clean and properly sealed to prevent contamination.

b. Personal Hygiene:
Wash hands frequently with soap and clean water, especially before eating or preparing food.

Maintain proper sanitation practices to minimize the spread of pathogens.

c. Avoiding Contaminated Water:
Avoid drinking water from questionable sources or those visibly contaminated.

Use reputable water sources and prioritize water safety during outdoor activities.

5. Seek Medical Attention:
If symptoms of a waterborne illness persist or worsen, seek medical attention promptly.

Report suspected waterborne illnesses to healthcare providers for proper diagnosis and treatment.

6. Emergency Preparedness:
Include waterborne illness treatment protocols in emergency preparedness plans.

Stock up on essential medications and supplies for managing waterborne illnesses in survival kits.

By understanding how to recognize, treat, and prevent waterborne illnesses, preppers can effectively manage health risks associated with contaminated water sources. Prioritize water safety measures and maintain vigilance in ensuring access to clean, safe drinking water during survival situations. Combining proper water treatment with hygiene practices and proactive healthcare strategies enhances overall readiness and resilience for preppers facing waterborne challenges in emergency scenarios.

CHAPTER EIGHT

Community Water Resources

Establishing Community Water Plans

In the context of survival and preparedness, establishing community water plans is vital for ensuring access to safe, reliable water resources during emergencies or disruptions. Community water planning involves collaborative efforts to assess water needs, identify sustainable water sources, implement treatment and purification methods, and develop resilience strategies. Here's a detailed guide on how preppers can establish effective community water plans:

1. Assessing Community Water Needs:
Conduct a thorough assessment of the community's water needs, considering factors such as population size, water consumption patterns, and existing water infrastructure.
Identify vulnerabilities and risks related to water supply, including potential disruptions or contamination events.

2. Identifying Water Sources:
Evaluate available water sources within or near the community, such as rivers, lakes, groundwater aquifers, rainwater harvesting systems, or community wells.

Prioritize reliable water sources that can sustain the community during emergencies or when traditional water supplies are compromised.

3. Developing Community Water Plans:
a. Water Quality Monitoring:
Establish regular water quality monitoring programs to ensure the safety and potability of community water sources.
Conduct testing for microbial contaminants, chemical pollutants, and physical parameters to identify potential risks.

b. Emergency Response Strategies:
Develop contingency plans and emergency response protocols for maintaining water supply during disasters, droughts, or other crises.
Establish alternative water supply options and backup systems to ensure continuous access to clean water in emergencies.

c. Water Treatment and Purification:
Implement water treatment and purification methods to remove contaminants and ensure safe drinking water.
Train community members on proper water treatment techniques and use of water purification technologies.

d. Infrastructure Development:
Invest in infrastructure projects to improve water storage, distribution, and sanitation facilities.

Upgrade aging water infrastructure and promote sustainable water management practices, including rainwater harvesting and water recycling.

e. Community Engagement and Education:
Engage community members through education programs on water conservation, hygiene practices, and emergency water preparedness.
 Foster partnerships with local stakeholders, government agencies, NGOs, and private sector entities to promote water stewardship and resilience.

4. Steps to Establishing Community Water Plans:
a. Forming a Water Committee:
Establish a dedicated water committee comprising community leaders, experts, and stakeholders to oversee water planning efforts.
 Ensure representation from diverse sectors to facilitate collaboration and decision-making.
b. Conducting Stakeholder Consultations:
 Engage community members in discussions and consultations to gather input, identify priorities, and build consensus around water planning objectives.
 Consider input from local residents, businesses, schools, healthcare providers, and emergency responders.

c. Drafting and Implementing the Water Plan:

Develop a comprehensive water plan outlining goals, objectives, and strategies for addressing water challenges and achieving water security.
 Allocate resources, set timelines, and assign responsibilities for plan implementation.

d. Monitoring and Evaluation:
 Regularly monitor progress, evaluate outcomes, and adjust strategies based on feedback and changing circumstances.
 Maintain flexibility to adapt water plans in response to evolving water conditions, community needs, or external factors.

5. Promoting Sustainability and Resilience:
Encourage water conservation practices and sustainable use of water resources to minimize waste and environmental impact.
 Foster resilience through diversified water supply sources, integrated water management approaches, and community-based adaptation strategies.

6. Collaboration and Networking:
 Collaborate with neighboring communities, regional organizations, and water management agencies to share resources, expertise, and best practices.
Establish partnerships for mutual support and collective action in addressing broader water challenges and promoting regional water security.

By establishing robust community water plans, preppers can enhance water resilience, promote sustainable water management practices, and ensure access to clean and safe drinking water for their communities during survival scenarios and emergencies. Investing in proactive water planning and preparedness fosters community empowerment, resilience, and overall readiness for water-related challenges in uncertain times.

Cooperation and Collaboration in Water Security

Ensuring water security in survival scenarios requires effective cooperation and collaboration among preppers, communities, and relevant stakeholders. By working together, preppers can optimize water management, enhance resilience, and secure access to clean and safe water resources. This guide provides detailed insights into the importance of cooperation and collaboration in water security for preppers:

1. Importance of Cooperation in Water Security:
Resource Sharing: Collaboration facilitates the sharing of resources, knowledge, and expertise related to water management, treatment, and conservation.

Enhanced Resilience: By working together, preppers can build resilience against water-related challenges such as shortages, contamination, or infrastructure disruptions.

Collective Problem-Solving: Cooperation enables collective problem-solving and innovation in addressing complex water security issues.

2. Strategies for Cooperation and Collaboration:
a. Establishing Water Committees:
Form community-based water committees comprising preppers, local residents, experts, and stakeholders.

Define roles, responsibilities, and decision-making processes to promote effective collaboration and accountability.

b. Sharing Information and Resources:
Exchange information, data, and best practices related to water management, treatment technologies, and emergency response strategies. Collaborate on water-related research, monitoring programs, and capacity-building initiatives.

c. Developing Mutual Aid Agreements:
Establish mutual aid agreements with neighboring communities or prepper's groups to provide support during water emergencies.

Coordinate mutual assistance for water supply, treatment, and distribution in times of crisis.

d. Engaging with Local Authorities and Agencies:
Foster partnerships with local government agencies, water utilities, and emergency responders to align water security efforts.

Participate in community planning processes to integrate water security considerations into local resilience strategies.

e. Promoting Water Conservation and Education:
Launch community outreach and education campaigns to raise awareness about water conservation, hygiene practices, and emergency preparedness.
Collaborate on initiatives to promote sustainable water use, rainwater harvesting, and water reuse.

f. Implementing Integrated Water Management Approaches:
Adopt integrated water management approaches that consider the entire water cycle, from source protection to wastewater treatment.
Collaborate on watershed management, groundwater recharge, and ecosystem restoration projects.

3. Building Networks and Partnerships:
Participate in regional or national networks focused on water security, emergency preparedness, and disaster response.
Forge partnerships with NGOs, academic institutions, and private sector entities to leverage resources and expertise.

4. Strengthening Community Resilience:
Strengthen community resilience by integrating water security into broader preparedness and sustainability initiatives.

Collaborate on multi-sectoral approaches that address interconnected challenges, such as food security, health, and infrastructure resilience.

5. Monitoring and Evaluation:
Establish monitoring systems to track progress, evaluate outcomes, and assess the effectiveness of collaborative water security efforts.
Use data-driven insights to inform adaptive management and continuous improvement in water management practices.

6. Advocating for Policy and Regulatory Support:
Advocate for policies and regulations that promote sustainable water management, water quality protection, and equitable access to water resources.
Collaborate with policymakers and advocacy groups to influence decision-making and promote water security at local, regional, and national levels.

By prioritizing cooperation and collaboration in water security efforts, preppers can enhance their resilience, optimize resource utilization, and ensure sustainable access to clean and safe water resources in survival scenarios. Embracing a collaborative approach fosters community empowerment, innovation, and collective action towards achieving water security goals in uncertain times.

Managing Shared Water Resources in Group Settings

Managing shared water resources in group settings is essential for preppers and communities to ensure sustainable access to clean water during survival scenarios. Effective management involves coordination, communication, and collaboration among group members to optimize water usage, prevent conflicts, and enhance water security.

1. Importance of Managing Shared Water Resources:

Equitable Access: Proper management ensures fair distribution of water, providing everyone in the group with access to this essential resource.

Efficient Usage: By coordinating usage, groups can minimize wastage and ensure water is used effectively to meet essential needs.

Conflict Prevention: Clear guidelines and agreements help prevent disputes over water allocation and usage, fostering cooperation within the group.

Sustainability: Collaborative management promotes sustainable practices, protecting water sources and ensuring long-term availability.

2. Establishing a Water Management Plan:
a. Identify Water Sources:

   Determine available water sources such as rivers, lakes, wells, or rainwater collection systems.

   Assess the reliability, quality, and capacity of each water source.

b. Conduct Needs Assessment:

Estimate the group's water requirements based on population size, daily usage, and specific needs (drinking, cooking, hygiene).

Consider factors like climate, seasonality, and potential emergencies when assessing water needs.

c. Develop Usage Guidelines:

Establish guidelines for water usage, including designated water collection times, rationing during shortages, and prioritization of essential needs.

Educate group members on responsible water usage and conservation practices.

d. Assign Responsibilities:

Delegate roles and responsibilities for water management tasks, such as water collection, purification, storage, and maintenance.

Ensure that responsibilities are clearly defined and understood by all group members.

3. Implementing Water Conservation Practices:

a. Educate Group Members:

Raise awareness about water conservation through training sessions, workshops, or informational materials.

Encourage behavioral changes, such as using water-efficient fixtures, repairing leaks, and minimizing non-essential water usage.

b. Adopt Sustainable Techniques:

Implement rainwater harvesting systems, greywater recycling, or water-saving irrigation methods to reduce reliance on external water sources.

Explore natural water purification methods, such as solar disinfection or biosand filtration, to treat water locally.

4. Monitoring and Maintenance:
a. Regular Inspections:

Conduct routine inspections of water sources, storage containers, and distribution systems to identify issues or potential contamination risks.

Address maintenance needs promptly to ensure the reliability and safety of water infrastructure.

b. Water Quality Testing:

Implement periodic water quality testing to monitor for microbial contaminants, chemical pollutants, and other potential risks.

Use portable testing kits or collaborate with local authorities for comprehensive water analysis.

5. Establishing Communication Protocols:
a. Coordination and Reporting:

Establish communication channels for sharing updates, reporting issues, and coordinating water-related activities within the group.

Designate a point of contact for water management inquiries and emergencies.

b. Conflict Resolution Mechanisms:

Develop protocols for resolving disputes or disagreements related to water usage, allocation, or management.

Encourage open dialogue and consensus-building to address conflicts constructively.

6. Building Resilience and Preparedness:
a. Emergency Preparedness:

Develop contingency plans for managing water shortages, disruptions, or contamination events.

Stockpile emergency supplies, including water purification tablets, filtration systems, and additional water storage containers.

b. Collaboration with External Partners:

Foster partnerships with neighboring groups, local authorities, or organizations to leverage resources, share best practices, and address common water challenges.

7. Review and Adaptation:
a. Continuous Improvement:

Conduct periodic reviews and evaluations of the water management plan to identify areas for improvement.

Incorporate feedback from group members and lessons learned from past experiences to enhance effectiveness and adaptability.

By implementing effective water management strategies in group settings, preppers can enhance their resilience, promote sustainability, and ensure access to clean water for survival and preparedness. Collaborative approaches foster cooperation, communication, and shared responsibility, strengthening the group's ability to manage and protect essential water resources in challenging circumstances.

CHAPTER NINE

Advanced Water Skills

In the realm of survival and preparedness, advanced water skills are essential for ensuring sustained hydration, health, and overall survival in challenging environments. While basic water procurement and treatment knowledge are foundational, advanced water skills involve specialized techniques and strategies that empower preppers to maximize water availability, efficiency, and safety in diverse and unpredictable conditions. This introduction provides an overview of why advanced water skills are crucial for preppers and survivalists.

1. The Critical Role of Water in Survival:
Water is a fundamental element for human survival, and its availability directly impacts our ability to thrive in any situation. Preppers recognize that access to clean and safe water is essential for hydration, food preparation, sanitation, and overall well-being, especially during emergencies or when living off-grid.

2. Importance of Advanced Water Skills:
While basic water procurement and treatment methods are fundamental, advanced water skills are necessary for addressing more complex

challenges and optimizing water resources. Advanced skills empower preppers to:

Identify and utilize alternative water sources beyond traditional options.

Implement advanced water purification and filtration techniques to remove contaminants effectively.

Conserve and manage water resources efficiently to ensure long-term sustainability.

Adapt to changing environmental conditions and unexpected challenges related to water availability and quality.

3. Challenges Addressed by Advanced Water Skills:

Water scarcity: Finding and utilizing unconventional water sources.

Water contamination: Safely treating and purifying water to remove harmful pollutants and pathogens.

Sustainability: Implementing practices to conserve and manage water resources effectively for extended periods.

Adaptability: Being prepared to handle unpredictable situations where traditional water sources may not be available or reliable.

4. Objectives of Advanced Water Skills Training:

The primary objectives of advancing water skills for preppers include:

Enhancing self-sufficiency and resilience in managing water-related emergencies.

Improving the ability to thrive in off-grid or remote settings by optimizing water resources.
Acquiring specialized knowledge and techniques to secure, treat, and store water safely and efficiently.

5. Scope of Advanced Water Skills:
Advanced water skills encompass a wide range of topics, including:
Rainwater harvesting and storage systems.
Wilderness water procurement and purification methods.
Advanced water filtration technologies.
Water conservation and sustainable management practices.
By investing in advanced water skills, preppers and survivalists can enhance their readiness and adaptability in challenging circumstances where water scarcity, contamination, or unpredictability are significant concerns. This guide will explore key aspects of advanced water skills to equip preppers with the knowledge and tools necessary for ensuring water security in survival situations

Water Purification from Non-Traditional Sources (Sea Water, Urine)

In survival scenarios, preppers may encounter situations where traditional freshwater sources are scarce or contaminated. Knowing how to purify water from non-traditional sources such as seawater and urine can be critical for ensuring hydration and survival. This guide explores detailed methods and considerations for purifying water from non-traditional sources in a water survival bible for preppers:

1. Purifying Sea Water:
Accessing drinkable water from seawater involves removing salt and other impurities through desalination methods. Preppers can use the following techniques to purify seawater:

a. Solar Desalination:
   Construct a solar still using a plastic sheet or tarp to collect evaporated freshwater from seawater.
   Place seawater in a container and cover with the plastic sheet, securing the edges.
   As the seawater evaporates under sunlight, freshwater condenses on the underside of the plastic and drips into a collection container.

b. Distillation:
   Boil seawater to create steam, then collect and condense the steam back into freshwater.
   Use a portable desalination device or DIY distillation setup to purify seawater through vapor condensation.

c. Reverse Osmosis:
   Use a portable reverse osmosis (RO) water purifier specifically designed for seawater desalination.

These devices use high-pressure membranes to filter out salts and contaminants, producing potable water.

2. Purifying Urine:

In emergency situations, urine can be purified into drinkable water through distillation or chemical treatment methods:

a. Solar Still for Urine Distillation:

Construct a solar still using a container filled with urine covered by a plastic sheet.

Place the still in direct sunlight, allowing urine vapor to condense on the plastic and collect as freshwater.

b. Chemical Treatment:

Use chemical disinfectants such as chlorine or iodine tablets to treat urine and kill harmful pathogens.

Follow manufacturer instructions for dosage and contact time to ensure effective purification.

c. Filtration and Purification:

Filter urine through a layered filtration system using materials like sand, charcoal, and fabric to remove impurities.

Follow up with chemical treatment or boiling to ensure microbial safety.

3. Considerations and Precautions:

a. Health Risks:

Purifying water from non-traditional sources may involve health risks due to high salt content (seawater) or contaminants (urine).

Ensure thorough purification and filtration to remove salts, pathogens, and other impurities.

b. Energy and Resources:
Desalination and distillation methods require energy inputs (e.g., sunlight, heat) and may be resource-intensive in survival settings.

c. Emergency Use Only:
Purifying water from non-traditional sources should be considered a last resort in emergency situations when no other freshwater sources are available.

4. Testing and Verification:
Use water quality testing kits to verify the effectiveness of purification methods and ensure water safety before consumption.
Monitor for signs of dehydration and seek medical attention if symptoms persist despite water purification efforts.

By understanding and applying these methods for purifying water from non-traditional sources like seawater and urine, preppers can expand their survival capabilities and increase their resilience in challenging environments. However, it's crucial to prioritize safety, proper techniques, and thorough purification to mitigate health risks associated with consuming water from non-traditional sources.

Water Recycling Advanced Filtration and Purification Techniques

In survival scenarios, having access to clean and safe drinking water is essential for maintaining health and ensuring survival. Advanced filtration and purification techniques enable preppers to treat water from various sources, removing contaminants and pathogens effectively. This guide explores detailed methods and considerations for advanced water filtration and purification in a water survival bible for preppers:

1. **Filtration Methods**:
a. **Activated Carbon Filtration**:
   Utilizes activated carbon (charcoal) to adsorb impurities, chemicals, and odors from water.
   Effective for removing chlorine, volatile organic compounds (VOCs), and some heavy metals.

b. **Ceramic Filtration:**
   Uses porous ceramic filters to physically trap bacteria, protozoa, and sediments from water.
   Often combined with activated carbon for enhanced filtration performance.

c. **Hollow Fiber Membrane Filtration**:
   Utilizes ultrafiltration membranes with microscopic pores to remove bacteria, protozoa, and particulates from water.
   Lightweight and portable, suitable for backpacking and emergency use.

d. Reverse Osmosis (RO) Filtration:

Uses semi-permeable membranes to remove salts, chemicals, and contaminants from water under pressure.

Effective for desalinating seawater and purifying brackish water.

2. Purification Methods:
a. Boiling:

Boil water vigorously for at least 1 minute (or longer at higher altitudes) to kill bacteria, viruses, and parasites.

Simple and effective method for emergency water purification.

b. Chemical Treatment:

Use chlorine dioxide tablets, iodine tablets, or household bleach (sodium hypochlorite) to disinfect water and kill pathogens.

Follow manufacturer instructions for proper dosage and contact time.

c. UV Water Purification:

Utilizes ultraviolet (UV) light to disrupt the DNA of bacteria, viruses, and protozoa, rendering them harmless.

Requires battery-operated UV purification devices or UV pen for portable use.

d. Ozonation:

Treat water with ozone gas to kill pathogens and improve taste by oxidizing contaminants.

Requires specialized ozone generators or ozonators for water purification.

3. Combination Filtration and Purification Systems:
a. Gravity-fed Water Filters:

Combine ceramic or hollow fiber membranes with activated carbon to provide comprehensive filtration and purification.

Suitable for treating large volumes of water for groups or families.

b. Portable Water Purification Devices:

Compact devices that integrate multiple filtration and purification methods (e.g., ceramic, carbon, UV) for versatile water treatment.

Ideal for backpacking, bug-out bags, or emergency kits.

4. Considerations and Precautions:
a. Water Quality Testing:

Use water quality testing kits to verify the effectiveness of filtration and purification methods.

Ensure treated water meets safety standards before consumption.

b. Maintenance and Replacement:

Regularly clean and maintain filtration devices to prevent clogging and ensure optimal performance.

Replace filter cartridges or components according to manufacturer recommendations.

c. Emergency Preparedness:

Include advanced filtration and purification systems in emergency preparedness kits to ensure access to clean water during disasters or crises.

5. Choosing the Right Method:

Consider factors such as water source, contaminants present, portability, and treatment capacity when selecting advanced filtration and purification techniques. Adapt methods based on specific survival scenarios and water quality requirements to optimize water safety and availability.

By mastering advanced filtration and purification techniques, preppers can enhance their resilience and self-reliance in securing clean drinking water from various sources in survival situations. Prioritize proper techniques, regular maintenance, and thorough water treatment to mitigate health risks and ensure water security for long-term survival.

Water recycling and Reclamation Methods

Water recycling and reclamation are essential strategies for preppers to maximize water resources, reduce waste, and maintain sustainable water supplies in survival scenarios. By implementing effective recycling and reclamation methods, preppers can extend their water reserves and improve overall water resilience. This guide explores detailed techniques and considerations for water recycling and reclamation in a water survival bible for preppers:

1. Importance of Water Recycling and Reclamation:

Water recycling and reclamation play a crucial role in water sustainability and self-sufficiency for preppers:

Conservation: Recycling water reduces reliance on finite water sources, conserving natural resources.

Resilience: Reclaimed water provides an alternative supply during water shortages or disruptions.

Waste Reduction: Recycling minimizes water wastage and promotes efficient resource utilization.

2. On-Site Water Recycling Methods:

a. Greywater Recycling:

Collects and treats wastewater from sinks, showers, and laundry for non-potable uses like toilet flushing, gardening, or laundry.

Use simple filtration, settling, and disinfection techniques to remove contaminants from greywater.

b. Rainwater Harvesting:

Collects rainwater from rooftops or surfaces into storage tanks or cisterns for non-potable uses.

Filter and disinfect rainwater for drinking or cooking purposes using appropriate treatment methods.

c. Condensation Harvesting:

Captures moisture from air through condensation using passive or active condensation devices.

Collect condensate for drinking or irrigation purposes after filtration and disinfection.

3. Advanced Water Reclamation Techniques:
a. Advanced Water Treatment (AWT):

Utilizes advanced treatment processes like membrane filtration, reverse osmosis, and UV disinfection to reclaim wastewater for potable reuse.

Removes contaminants and pathogens to meet drinking water standards.

b. Desalination:

Converts seawater or brackish water into freshwater through desalination processes like reverse osmosis, distillation, or electrodialysis.

Provides alternative freshwater sources in coastal or arid regions.

c. Aquaponics and Hydroponics:

Integrates water recycling with food production using aquaponics (fish and plant cultivation) or hydroponics (soil-less plant cultivation).

Recycles nutrient-rich water from aquaculture or hydroponic systems for plant irrigation, reducing water waste.

4. Considerations and Precautions:
a. Water Quality Assurance:
Implement robust filtration, disinfection, and monitoring systems to ensure reclaimed water meets safety and quality standards.

Regularly test recycled water for contaminants and pathogens.

b. System Maintenance:
Maintain recycling systems by cleaning filters, replacing components, and addressing maintenance issues promptly to prevent system failures.

Ensure proper operation and functionality of water recycling infrastructure.

c. Regulatory Compliance:
Familiarize yourself with local regulations and guidelines for water recycling and reuse to ensure compliance with health and environmental standards.

Obtain necessary permits or approvals for implementing advanced water reclamation systems.

5. Integrating Water Recycling into Preparedness Plans:
Include water recycling and reclamation strategies in prepping and survival plans:

Designate space for rainwater harvesting, greywater systems, or aquaponics/hydroponics setups.
Educate group members on water recycling practices and promote sustainable water use habits.

By incorporating water recycling and reclamation methods into survival preparations, preppers can enhance their self-reliance, reduce dependency on external water sources, and improve overall water sustainability for long-term survival and resilience. Prioritize proper treatment, maintenance, and regulatory compliance to maximize the benefits of water recycling in survival scenarios.

CHAPTER TEN

Planning for Long-Term Water Security

Developing Sustainable Water Solutions

In survival scenarios, access to clean and sustainable water sources is crucial for preppers to ensure hydration, sanitation, and overall well-being. Developing sustainable water solutions involves implementing strategies that maximize water availability, minimize wastage, and promote long-term resilience. This guide explores detailed methods and considerations for developing sustainable water solutions in a water survival bible for preppers:

1. Assessing Water Needs:
Understanding water requirements based on factors like climate, activity level, and health is fundamental to developing sustainable water solutions. Calculate daily water needs per person and consider additional requirements for food preparation, hygiene, and sanitation.

2. Water Storage Techniques:

Effective water storage is essential for preppers to maintain a reliable water supply. Consider the following techniques:

Rainwater Harvesting: Collect rainwater from rooftops using gutters and downspouts into storage tanks or cisterns.

Well water: Regularly maintain wells and establish new well systems to access groundwater.

Surface Water Collection: Identify and utilize safe surface water sources such as rivers, lakes, or streams for water storage.

3. Water Filtration and Purification:

Implement robust water filtration and purification methods to ensure water safety:

Filtration: Use ceramic filters, activated carbon, or hollow fiber membranes to remove sediments and contaminants from water.

Purification: Boil water, use chemical disinfectants (e.g., chlorine tablets, iodine), or employ UV light treatment to kill harmful pathogens.

4. Implementing Water Conservation Practices:

Promote water conservation to minimize wastage and extend water availability:

Greywater Recycling: Reuse water from sinks, showers, and laundry for non-potable purposes like irrigation or toilet flushing.

Efficient Irrigation: Use drip irrigation systems or water-saving techniques in gardening to optimize water usage.
Minimizing Leaks and Wastage: Regularly inspect and repair plumbing leaks to prevent water loss.

5. Sustainable Water Sources:
Explore alternative and sustainable water sources for long-term resilience:

Desalination: Investigate desalination methods to convert seawater into potable freshwater, such as reverse osmosis or solar distillation.
Aquaponics and Hydroponics: Integrate water recycling with food production using aquaponic or hydroponic systems.

6. Collaborative Water Solutions:
Engage in community-based water initiatives to enhance water security:
Water Co-ops: Collaborate with neighbors to share resources and establish community water management systems.
Collective Conservation Efforts: Educate and involve local communities in water conservation projects to promote sustainability.

7. Emergency Preparedness and Adaptation:
Integrate water solutions into comprehensive emergency preparedness plans:

Monitor and Maintain: Regularly monitor water systems and storage facilities to ensure functionality during emergencies.
Adapt to Changing Conditions: Stay flexible and adapt water strategies based on evolving environmental or crisis situations.

8. Education and Training:
Provide education and training on water conservation and sustainable practices to enhance community resilience and self-reliance.

By incorporating these detailed methods and considerations into their survival preparations, preppers can develop sustainable water solutions that ensure long-term water availability, promote self-sufficiency, and enhance resilience in challenging circumstances. Prioritize water security as a fundamental aspect of survival planning, and continually assess and refine water management strategies to adapt to changing conditions and needs.

Permaculture Techniques for Water Management

Permaculture offers a holistic approach to sustainable agriculture and land management, emphasizing the integration of natural systems to maximize efficiency and minimize waste. Incorporating permaculture techniques into water management strategies can significantly enhance water security and resilience for preppers in survival scenarios. This guide explores detailed permaculture techniques for water management in a water survival bible for preppers:

1. Swales and Contouring:
Swales are shallow ditches dug along contour lines to capture and slow down water runoff. This technique helps to recharge groundwater, reduce soil erosion, and maximize water infiltration into the soil. Preppers can implement swales on their property to channel rainwater and promote water retention in the landscape.

2. Keyline Design:
Keyline design involves identifying and cultivating the "keyline" contour of the land to manage water flow effectively. By contouring the land along these keylines, preppers can direct water to specific areas, such as gardens or water storage basins, optimizing water distribution and utilization.

3. Mulching and Soil Improvement:
Mulching with organic materials like straw, leaves, or wood chips helps retain soil moisture, reduce evaporation, and improve soil structure. Healthy soil with high organic matter content can absorb and hold water more effectively, reducing the need for irrigation and enhancing plant resilience during dry periods.

4. Rainwater Harvesting Earthworks:
Constructing earthworks such as ponds, dams, and swales can capture and store rainwater for irrigation and domestic use. Preppers can design and implement earthworks based on permaculture principles to enhance water availability and create resilient water features on their property.

5. Plant Guilds and Water-Conserving Plants:
Designing plant guilds—combinations of mutually beneficial plants—can enhance water efficiency in gardens. Planting drought-tolerant species and using companion planting techniques can reduce water requirements and create resilient ecosystems that thrive with minimal irrigation.

6. Greywater Recycling and Treatment:
Recycling greywater (wastewater from sinks, showers, and laundry) for irrigation can minimize freshwater usage in gardens. Preppers can set up simple greywater systems that filter and treat wastewater using natural processes like biofiltration or constructed wetlands before redirecting it to water plants.

7. Water-Efficient Design and Infrastructure:
Integrate water-efficient design principles into infrastructure and landscaping:
Use permeable paving materials to reduce runoff and allow water to infiltrate into the ground.

Install rainwater harvesting systems with guttering, downspouts, and storage tanks to collect roof runoff.

8. Holistic Water Planning:
Take a holistic approach to water management by considering the entire water cycle:
Assessing water needs and sources on-site.
Designing integrated water systems that mimic natural patterns and cycles.
Implementing adaptive management practices that respond to changing environmental conditions.

9. Education and Community Engagement:
Share knowledge and practices with local communities to promote water-wise landscaping and sustainable water management practices. Collaborate with neighbors to implement collective water solutions that enhance resilience and self-sufficiency.

By incorporating permaculture techniques for water management into their survival preparations, preppers can create self-sustaining landscapes that maximize water efficiency, enhance biodiversity, and promote long-term water security. Prioritize regenerative practices that work with nature to conserve water resources and build resilient ecosystems capable of withstanding diverse environmental challenges.

Integration of Water Security into Overall Preparedness Plans

Water security is a critical component of any comprehensive preparedness plan for preppers. Ensuring access to clean and reliable water sources is essential for survival in emergency situations. This guide explores detailed strategies and considerations for integrating water security into overall preparedness plans in a water survival bible for preppers:

1. Water Assessment and Inventory:
Start by assessing current water sources and storage capacity:
 Identify available water sources (e.g., wells, rainwater collection, surface water).
 Calculate water storage capacity and estimate daily water needs for each person in the household.

2. Diverse Water Sources:
Diversify water sources to enhance resilience:
Utilize multiple sources such as well water, rainwater harvesting, and surface water (rivers, lakes).
 Consider alternative sources like desalination techniques for coastal regions.

3. Water Storage and Treatment:
Optimize water storage and treatment systems:

Invest in quality water storage containers (tanks, barrels) with sufficient capacity.
 Implement effective water treatment methods (filtration, boiling, chemical treatment) to ensure water safety.

4. Emergency Water Supply:
Establish emergency water supply reserves:
Maintain a minimum supply of stored water (e.g., 3-day, 7-day, or longer) for each household member.
Rotate water storage periodically to ensure freshness and quality.

5. Greywater Recycling and Conservation:
Implement greywater recycling and conservation practices:
Install greywater systems to recycle wastewater for non-potable uses (e.g., irrigation, toilet flushing).
Promote water-saving habits (e.g., fixing leaks, using water-efficient appliances) to minimize water consumption.

6. Rainwater Harvesting and Management:
Maximize rainwater harvesting for sustainable water supply:
Set up rainwater collection systems with guttering and storage tanks to capture roof runoff.
 Use rain barrels or cisterns to store harvested rainwater for domestic use.

7. Community Engagement and Collaboration:

Engage with local communities to enhance water security:
Collaborate with neighbors to share resources and implement collective water solutions (e.g., community wells, water co-ops).
Organize workshops or training sessions on water conservation and emergency preparedness.

8. Monitoring and Maintenance:
Regularly monitor and maintain water systems:
Conduct routine inspections of water storage containers, filtration systems, and plumbing infrastructure.
Replace filters, disinfect storage tanks, and address maintenance issues promptly.

9. Education and Training:
Provide education and training on water security and emergency water management:
Educate family members on water conservation practices and emergency water treatment techniques.
Train individuals in the use of water filtration devices, purification methods, and alternative water sources.

10. Adaptation and Resilience:
Adapt water security strategies to changing conditions and risks:

Develop contingency plans for water shortages or contamination events.

Stay informed about local water regulations, weather patterns, and environmental changes that may impact water availability.

11. Continuous Improvement:
Continuously evaluate and improve water security measures:
Seek feedback from community members and adapt strategies based on lessons learned.
Invest in technology and innovations that enhance water efficiency and sustainability.

By integrating water security into overall preparedness plans, preppers can enhance resilience, self-sufficiency, and readiness to manage water-related challenges during emergencies or survival scenarios. Prioritize proactive water management strategies, community collaboration, and ongoing education to ensure sustained access to clean and reliable water for long-term survival.

CONCLUSION

Final Thoughts on Water Survival Preparedness

In the pursuit of comprehensive survival preparedness, water security stands as one of the most vital and fundamental aspects. As you conclude your journey through water survival preparedness, it's essential to reflect on key considerations and takeaways to ensure sustained resilience and readiness in the face of adversity:

1. Water is Life: Recognize that water is not just a resource but a life-sustaining necessity. Prioritize water security as a foundational element of your preparedness strategy.

2. Diverse Water Sources: Embrace diversity in water sources to enhance resilience. Incorporate multiple sources such as wells, rainwater harvesting, and surface water to mitigate risks and ensure access during various scenarios.

3. Storage and Treatment: Invest in reliable water storage containers and effective treatment methods. Properly store and treat water to maintain safety and quality over time.

4. Conservation and Recycling: Implement water conservation practices and explore greywater recycling systems. Minimize wastage and maximize efficiency to extend water availability.

5. Community Collaboration: Engage with local communities to strengthen water security collectively. Collaborate with neighbors, share resources, and promote mutual support in managing water challenges.

6. Adaptability and Innovation: Stay adaptable and open to innovation. Continuously assess and refine water management strategies based on evolving conditions, emerging technologies, and lessons learned.

7. Education and Training: Foster a culture of education and training on water survival preparedness. Equip yourself and others with essential knowledge and skills to confidently navigate water-related challenges.

8. Long-Term Sustainability: Strive for long-term sustainability in water management. Integrate permaculture principles, rainwater harvesting, and ecological design to create resilient water systems.

9. Monitoring and Maintenance: Regularly monitor and maintain water systems to ensure reliability. Conduct routine inspections, address maintenance issues promptly, and stay proactive in safeguarding water sources.

10. Hope and Resilience: Maintain hope and resilience in the face of uncertainty. Water survival preparedness is not just about physical resources but also about mental fortitude and adaptability.

In closing, water survival preparedness is a journey that requires dedication, foresight, and continuous effort. By embracing holistic approaches, fostering community connections, and prioritizing sustainable practices, you can navigate water-related challenges with confidence and resilience. Remember, water is the essence of life, and with thoughtful preparation, you can secure this vital resource for yourself and your loved ones in any situation.
Stay informed, stay prepared, and stay resilient. Your commitment to water survival preparedness will serve you well in the pursuit of a resilient and self-sufficient lifestyle.

Encouragement for Ongoing Learning and Adaptation

As you embark on your journey towards water survival preparedness, remember that continuous learning and adaptation are key to staying resilient in the face of evolving challenges. Here are some words of encouragement to guide you in your pursuit of sustainable water solutions:

1. Stay Informed and Updated; Keep yourself updated with the latest advancements in water purification techniques, sustainable practices, and emergency preparedness strategies. Follow reputable sources, join online forums, and participate in workshops or training sessions to expand your knowledge base.

2. Embrace Flexibility and Innovation: Be open to exploring new ideas and innovative approaches to water conservation and management. Experiment with different techniques and adapt them to suit your unique circumstances and environment.

3. Build Community Resilience: Collaborate with like-minded individuals and communities to share knowledge, resources, and experiences. Together, you can develop collective solutions and strengthen each other's resilience during times of crisis.

4. Practice Regular Maintenance: Regularly inspect and maintain your water storage systems, purification tools, and harvesting infrastructure. This proactive approach will ensure that your water sources remain reliable and functional when you need them most.

5. Prioritize Sustainability: Strive for sustainable water practices that minimize environmental impact and promote long-term resource conservation. Implement water-saving habits in your daily life and encourage others to do the same.

6. Adapt to Changing Conditions: Recognize that water availability and needs can change over time due to various factors like climate variability or population growth. Stay adaptable and be prepared to adjust your strategies accordingly.

7. Celebrate Progress and Milestones: Acknowledge your achievements, no matter how small. Celebrate successful water harvesting, purification experiments, or community initiatives to stay motivated and inspired.

Remember, the journey towards water survival preparedness is not just about acquiring skills—it's about cultivating a mindset of resilience, adaptability, and continuous improvement. By embracing ongoing learning and adaptation, you will be better equipped to face whatever challenges the future may hold. Keep striving, keep learning, and keep innovating towards a more sustainable and secure water future.

www.ingramcontent.com/pod-product-compliance
Lightning Source LLC
Chambersburg PA
CBHW052205220526
45471CB00004B/1831